ASCENT
CENTER FOR TECHNICAL KNOWLEDGE

Autodesk® Inventor® 2017 (R1) iLogic

Student Guide
Mixed Units - 1st Edition

AUTODESK.
Authorized Publisher

ASCENT - Center for Technical Knowledge®
Autodesk® Inventor® 2017 (R1)
iLogic
1st Edition

Prepared and produced by:

ASCENT Center for Technical Knowledge
630 Peter Jefferson Parkway, Suite 175
Charlottesville, VA 22911

866-527-2368
www.ASCENTed.com

Lead Contributor: Jennifer MacMillan

ASCENT - Center for Technical Knowledge is a division of Rand Worldwide, Inc., providing custom developed knowledge products and services for leading engineering software applications. ASCENT is focused on specializing in the creation of education programs that incorporate the best of classroom learning and technology-based training offerings.

We welcome any comments you may have regarding this student guide, or any of our products. To contact us please email: feedback@ASCENTed.com.

Contents

Preface

The *Autodesk® Inventor® 2017 (R1) iLogic* student guide teaches students to use the iLogic functionality that exists in the Autodesk® Inventor® 2017 (R1) software. In this practice-intensive curriculum, students acquire the knowledge required to use iLogic to automate Autodesk Inventor designs.

In this student guide, students learn how iLogic functionality furthers the use of parameters in a model by adding an additional layer of intelligence. By setting criteria in the form of established rules they learn how to capture design intent, enabling them to automate the design workflow to meet various design scenarios in part, assembly, and drawing files.

Topics Covered

- iLogic functionality overview.

- iLogic workflow overview.

- Review of model and user-defined parameters, and equations and their importance in iLogic.

- Understanding the iLogic interface components (iLogic Panel, Edit Rule dialog box, and iLogic Browser).

- Rule creation workflow for Autodesk Inventor parts and assemblies.

- Using variations of conditional statements in an iLogic rule.

- Accessing and incorporating the various function types into an iLogic part, assembly, or drawing file rule.

- Event Triggers and iTriggers.

- Creating Forms to create a custom user interface for an iLogic rule.

Note on Software Setup

This student guide assumes a standard installation of the software using the default preferences during installation. Lectures and practices use the standard software templates and default options for the Content Libraries.

Students and Educators can Access Free Autodesk Software and Resources

Autodesk challenges you to get started with free educational licenses for professional software and creativity apps used by millions of architects, engineers, designers, and hobbyists today. Bring Autodesk software into your classroom, studio, or workshop to learn, teach, and explore real-world design challenges the way professionals do.

Get started today - register at the Autodesk Education Community and download one of the many Autodesk software applications available.

Visit www.autodesk.com/joinedu/

Note: Free products are subject to the terms and conditions of the end-user license and services agreement that accompanies the software. The software is for personal use for education purposes and is not intended for classroom or lab use.

Lead Contributor: Jennifer MacMillan

With a dedication for engineering and education, Jennifer has spent over 20 years at ASCENT managing courseware development for various CAD products. Trained in Instructional Design, Jennifer uses her skills to develop instructor-led and web-based training products as well as knowledge profiling tools.

Jennifer has achieved the Autodesk Certified Professional certification for Inventor and is also recognized as an Autodesk Certified Instructor (ACI). She enjoys teaching the training courses that she authors and is also very skilled in providing technical support to end-users.

Jennifer holds a Bachelor of Engineering Degree as well as a Bachelor of Science in Mathematics from Dalhousie University, Nova Scotia, Canada.

Jennifer MacMillan has been the Lead Contributor for *Autodesk Inventor iLogic* since its initial release in 2013.

In this Guide

The following images highlight some of the features that can be found in this Student Guide.

Practice Files

The Practice Files page tells you how to download and install the practice files that are provided with this student guide.

FTP link for practice files

Chapters

Each chapter begins with a brief introduction and a list of the chapter's Learning Objectives.

Learning Objectives for the chapter

Side notes

Side notes are hints or additional information for the current topic.

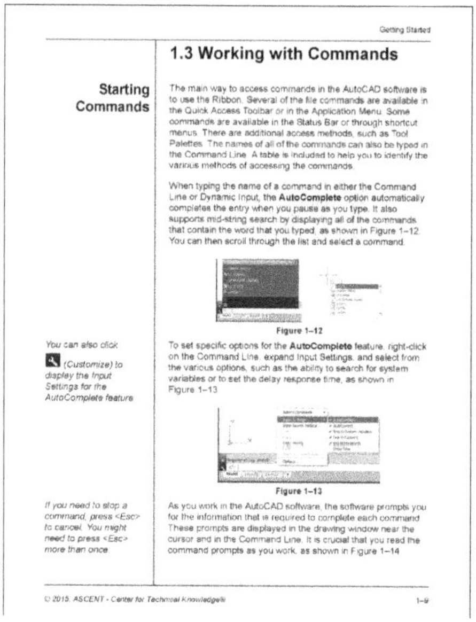

Getting Started

1.3 Working with Commands

Starting Commands

The main way to access commands in the AutoCAD software is to use the Ribbon. Several of the file commands are available in the Quick Access Toolbar or in the Application Menu. Some commands are available in the Status Bar or through shortcut menus. There are additional access methods, such as Tool Palettes. The names of all of the commands can also be typed in the Command Line. A table is included to help you to identify the various methods of accessing the commands.

When typing the name of a command in either the Command Line or Dynamic Input, the **AutoComplete** option automatically completes the entry when you pause as you type. It also supports mid-string search by displaying all of the commands that contain the word that you typed, as shown in Figure 1–12. You can then scroll through the list and select a command.

Figure 1–12

You can also click (Customize) to display the Input Settings for the AutoComplete feature.

To set specific options for the **AutoComplete** feature, right-click on the Command Line, expand Input Settings, and select from the various options, such as the ability to search for system variables or to set the delay response time, as shown in Figure 1–13.

Figure 1–13

If you need to stop a command, press <Esc> to cancel. You might need to press <Esc> more than once.

As you work in the AutoCAD software, the software prompts you for the information that is required to complete each command. These prompts are displayed in the drawing window near the cursor and in the Command Line. It is crucial that you read the command prompts as you work, as shown in Figure 1–14.

© 2015, ASCENT - Center for Technical Knowledge® 1–9

Instructional Content

Each chapter is split into a series of sections of instructional content on specific topics. These lectures include the descriptions, step-by-step procedures, figures, hints, and information you need to achieve the chapter's Learning Objectives.

Practice Objectives

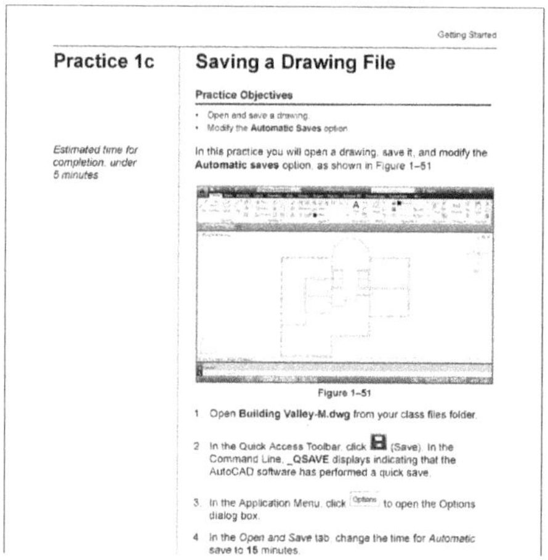

Getting Started

Practice 1c **Saving a Drawing File**

Practice Objectives

- Open and save a drawing.
- Modify the **Automatic Saves** option.

Estimated time for completion: under 5 minutes

In this practice you will open a drawing, save it, and modify the **Automatic saves** option, as shown in Figure 1–51.

Figure 1–51

1. Open **Building Valley-M.dwg** from your class files folder.
2. In the Quick Access Toolbar, click (Save). In the Command Line, **_QSAVE** displays indicating that the AutoCAD software has performed a quick save.
3. In the Application Menu, click Options to open the Options dialog box.
4. In the Open and Save tab, change the time for *Automatic save* to **15** minutes.

Practices

Practices enable you to use the software to perform a hands-on review of a topic.

Some practices require you to use prepared practice files, which can be downloaded from the link found on the Practice Files page.

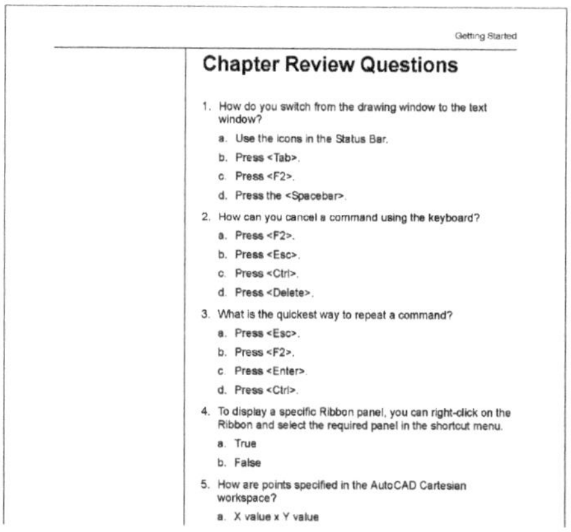

Getting Started

Chapter Review Questions

1. How do you switch from the drawing window to the text window?
   a. Use the icons in the Status Bar.
   b. Press <Tab>.
   c. Press <F2>.
   d. Press the <Spacebar>.
2. How can you cancel a command using the keyboard?
   a. Press <F2>.
   b. Press <Esc>.
   c. Press <Ctrl>.
   d. Press <Delete>.
3. What is the quickest way to repeat a command?
   a. Press <Esc>.
   b. Press <F2>.
   c. Press <Enter>.
   d. Press <Ctrl>.
4. To display a specific Ribbon panel, you can right-click on the Ribbon and select the required panel in the shortcut menu.
   a. True
   b. False
5. How are points specified in the AutoCAD Cartesian workspace?
   a. X value x Y value

Chapter Review Questions

Chapter review questions, located at the end of each chapter, enable you to review the key concepts and learning objectives of the chapter.

Getting Started

Command Summary

The following is a list of the commands that are used in this chapter, including details on how to access the command using the software's Ribbon, toolbars, or keyboard commands.

Button	Command	Location
	Close	• Drawing Window • Application Menu • Command Prompt: close
	Close Current Drawing	• Application Menu
	Close All Drawings	• Application Menu
NA	Dynamic Input	• Status Bar: expand Customization
Exit Autodesk AutoCAD	Exit AutoCAD	• Application Menu
	Open	• Quick Access Toolbar • Application Menu • Command Prompt: open, <Ctrl>+<O>
	Open Documents	• Application Menu
Options	Options	• Application Menu • Shortcut Menu: Options
	Pan	• Navigation Bar • Shortcut Menu: Pan • Command Prompt: pan or P
	Recent Documents	• Application Menu
	Save	• Quick Access Toolbar • Application Menu • Command Prompt: qsave, <Ctrl>+<S>
	Save As	• Quick Access Toolbar • Application Menu • Command Prompt: save
	Zoom Realtime	• Navigation Bar: Zoom Realtime • Shortcut Menu: Zoom

Command Summary

The Command Summary is located at the end of each chapter. It contains a list of the software commands that are used throughout the chapter, and provides information on where the command is found in the software.

Practice Files

To download the practice files for this student guide, use the following steps:

1. Type the URL shown below into the address bar of your Internet browser. The URL must be typed **exactly as shown**. If you are using an ASCENT ebook, you can click on the link to download the file.

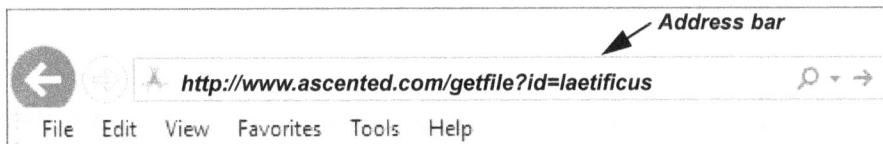

Address bar

http://www.ascented.com/getfile?id=laetificus

File   Edit   View   Favorites   Tools   Help

2. Press <Enter> to download the .ZIP file that contains the Practice Files.

3. Once the download is complete, unzip the file to a local folder. The unzipped file contains an .EXE file.

4. Double-click on the .EXE file and follow the instructions to automatically install the Practice Files on the C:\ drive of your computer.

    Do not change the location in which the Practice Files folder is installed. Doing so can cause errors when completing the practices in this student guide.

http://www.ascented.com/getfile?id=laetificus

Stay Informed!

Interested in receiving information about upcoming promotional offers, educational events, invitations to complimentary webcasts, and discounts? If so, please visit:

www.ASCENTed.com/updates/

Help us improve our product by completing the following survey:

www.ASCENTed.com/feedback

You can also contact us at: *feedback@ASCENTed.com*

Introduction to iLogic

The Autodesk® Inventor® iLogic functionality provides an easy-to-use interface that enables you to automate designs by creating and manipulating rules that drive the model's geometry. As opposed to making parameter changes in a single model, the use of iLogic enables you to set the controls for parameter change to build design intent into a model.

Learning Objectives in this Chapter

- Understand the various design automation tools in the Autodesk Inventor software.
- Understand how the iLogic functionality can be incorporated into a model to use parameters and equations to build design intent into a model.
- Describe the workflow for incorporating iLogic automation into an Autodesk Inventor file.

1.1 Design Automation Overview

Several different methodologies can be used for design automation. These range from out-of-the-box Autodesk Inventor solutions to highly customized and programmed solutions. The progression of these solutions is shown in Figure 1–1. As you progress through the options, the investment in training or customized programming increases. The key is to find a solution that balances a justifiable investment with the requirements for the solution.

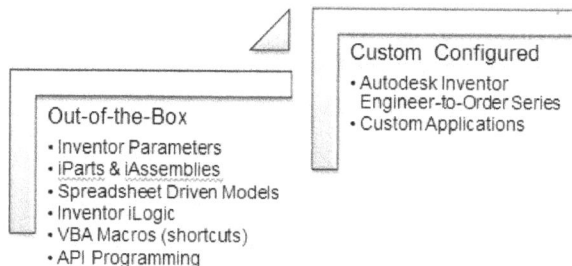

Custom Configured
- Autodesk Inventor Engineer-to-Order Series
- Custom Applications

Out-of-the-Box
- Inventor Parameters
- iParts & iAssemblies
- Spreadsheet Driven Models
- Inventor iLogic
- VBA Macros (shortcuts)
- API Programming

Figure 1–1

Out-of-the-box

A number of tools in the Autodesk Inventor software can be used to incorporate intelligence into your designs. The progression of these solutions in terms of complexity is shown in Figure 1–2.

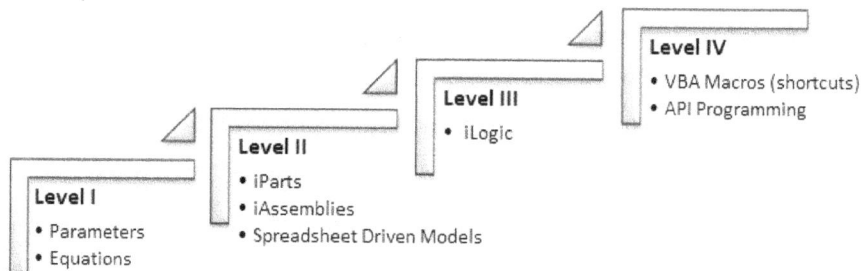

Level IV
- VBA Macros (shortcuts)
- API Programming

Level III
- iLogic

Level II
- iParts
- iAssemblies
- Spreadsheet Driven Models

Level I
- Parameters
- Equations

Figure 1–2

- Parameters and equations can be used to relate dimensions or features to one another. The relationships that are established build intelligence into your design so that the basic design intent is captured.

- The Autodesk Inventor iPart and iAssembly functionality enables you to create variations in your part designs quickly and easily. This provides an alternative to recreating the same model repeatedly while varying specific parameters in the model. This customization is accomplished by assigning variable model parameters to a table in which values are entered to create a specific variation.

- Parameters can be controlled directly in the software. However, a further level of customization enables you to control the parameter values using a spreadsheet. The spreadsheet can be embedded or linked to an Autodesk Inventor file to control the parameter values. This is ideal for users that might not have access to the Autodesk Inventor software or have limited knowledge of the software.

- Using iLogic functionality in the Autodesk Inventor software enables you to incorporate a rules-based design into your model that captures and automates an intelligent design process. Using the basic iLogic functions requires little or no programming knowledge, but programming experience can help with the advanced functions. This functionality is the focus of this student guide.

- Programming VBA Macros or Application Programming Interfaces (API), or incorporating custom add-ins for design automation can be done using the out-of-the-box Autodesk Inventor functionality. This level of customization requires programming knowledge.

Many users do not have API experience or a background with VB and VBA programming that is required for the top level of out-of-the-box customization. Therefore, iLogic, with its intuitive interface, is an excellent choice for design automation.

Custom Configured

At the top end of the spectrum, standalone tools can be programmed that are highly customized and oriented toward solving a specific business or design need. For example, an engineer-to-order system or sales configurator tool might be required to limit the contact points in a sales cycle. This solution generally requires programming using the Autodesk® Inventor® Engineer-to-Order Series or other custom-built applications with considerable investment.

1.2 iLogic Overview

iLogic can also be used in the Assembly, Drawing, and Sheet Metal environments.

iLogic functionality takes the use of parameters and equations in a model one step further by adding an additional layer of intelligence in your design. iLogic enables you to set the criteria in the form of established rules that capture design intent, enabling you to reuse designs to suit various design scenarios. The rules are all incorporated as part of the digital prototype definition. The rules that define the design intent control the model and automate the design workflow to ensure that the model reacts correctly. iLogic rules are based on conditional statements and functions. Automation functions have been embedded in the rule creation dialog box enabling you to define the rules that can accomplish many different results in a design. No programming knowledge is required to use the basic iLogic functions, but programming experience can be an asset when using the advanced functions.

iLogic is VB-based. Therefore, visual basic code can be used.

In the example shown in Figure 1–3, iLogic is used to control the thickness of the connecting rod, based on its length. The rule that controls the geometry indicates the range of values that are acceptable and what thickness value is to be used.

```
If length >= 60 And length < 80 Then
Parameter("thickness") = 1.5

Else If length >= 80 And length <= 100 Then
Parameter("thickness") = 3.0

Else If length < 60 Or length > 100 Then
MessageBox.Show("Please enter a value between 60-100 mm.", "Invalid Entry")

End If
```

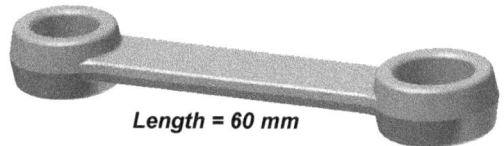

Length = 60 mm

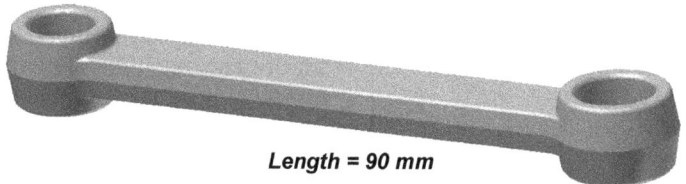

Length = 90 mm

Figure 1–3

iLogic Functionality

Some things that can be accomplished using iLogic rules include the following:

- Control model and user parameter values to ensure that the specifications and standards are met. Supports string (Text), boolean (True/False), and numeric parameters.

- Activate part and assembly features, assembly components, or assembly constraints based on conditional statements.

- Perform multiple operations based on a single user input.

- Perform checks (iProperties, dimensions, etc.) in the model for design situations.

- Update material and iProperty information in the model.

- Read document information (filename, path, extension, etc.).

- Measure entities in the model.

- Provide customized feedback based on specified conditions.

- Drive iFeature, iPart, and iAssembly configurations.

- Incorporate the execution of other rules in a parent rule.

- Interface directly with a predefined form to assign parameter values.

- Control drawing size, borders, and title block information based on user entry.

- Control view location and size or suppression in a drawing.

iLogic Examples

The assemblies shown in Figure 1–4 have been automated using iLogic. Custom forms have been created to easily enter the required data that drives all of the iLogic rules.

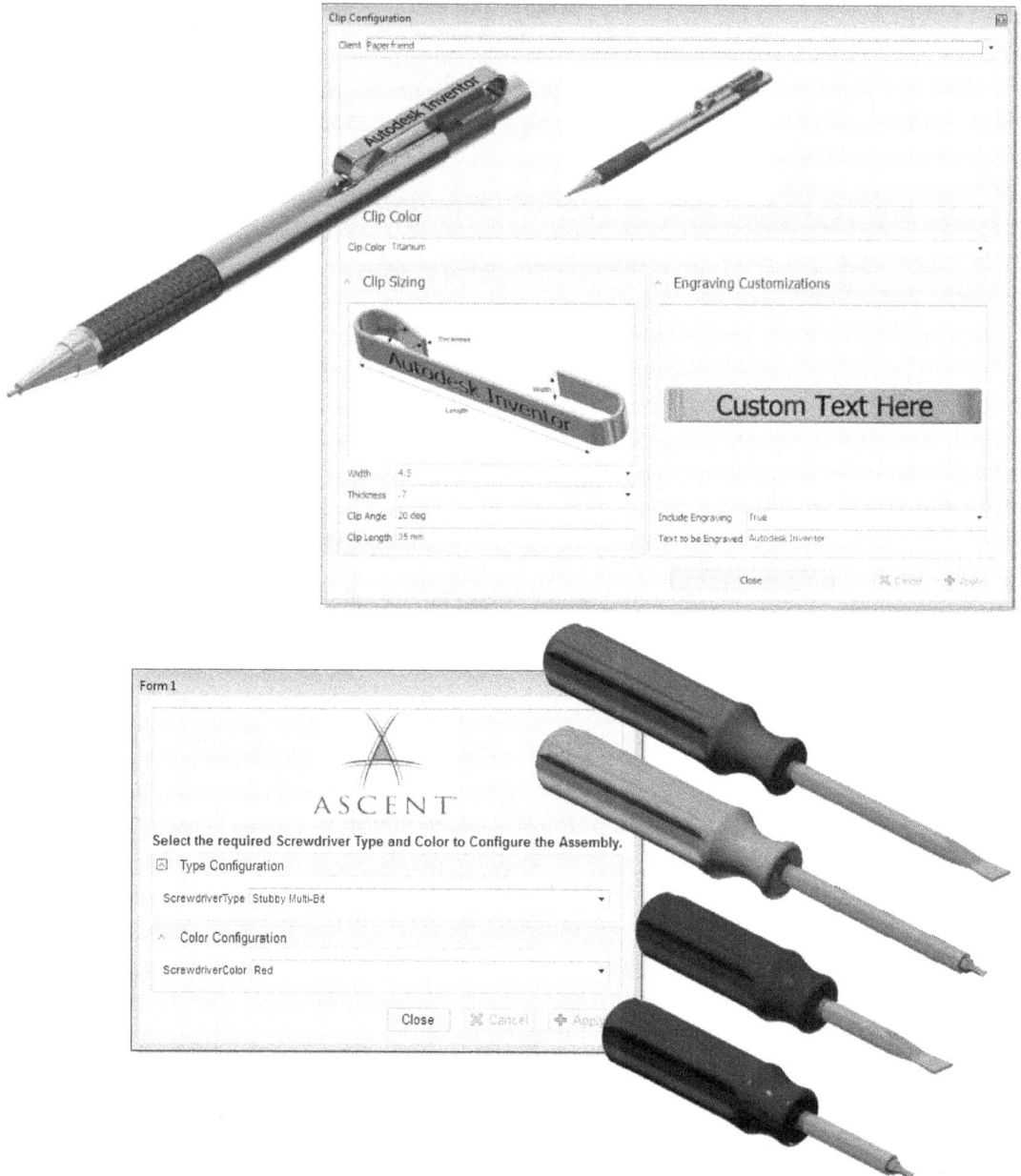

Figure 1–4

1.3 iLogic Workflow

As with most functionality in the Autodesk Inventor software, the use of iLogic also follows a workflow. This workflow can be broken into four steps.

General Steps

Use the following general steps to create iLogic rules:

1. Prepare the Model/Drawing.
2. Rule Creation.
3. Set rule triggers.
4. Create and edit rules, as necessary.

Figure 1–5 highlights the steps graphically. Additional in-depth information is included for each of these steps as you progress through this student guide.

Prepare the Model/Drawing → Rule Creation → Set Rule Triggers → Create and Edit Rules, as necessary

Figure 1–5

Step 1 - Prepare the Model/Drawing.

Before a rule can be created, all of the required dimensions, parameters, and equations must already have been added to the model. Rules are written based on this information and without it, the rules are not tied to the model geometry. Therefore, when you design a model, you should always consider the final design intent.

- Verify that feature dimensions capture the model's geometric intent, ensure that any required user parameters (Numeric, Text, and True/False) are included, and that they update if changes are required.

In the case of automating a drawing, a drawing must exist that contains all of the details required to communicate the design. iLogic rules in a drawing can control sheet sizes, title blocks and borders, view positioning, scaling, and suppression.

Step 2 - Rule Creation.

An iLogic rule is used to control the parameters, features, or components beyond what the user defined parameters and equations can do in the Autodesk Inventor software.

- iLogic rules are based on conditional statements and functions that capture the design intent, enabling you to reuse designs to meet various design scenarios.

- A substantial and varied list of functions are provided that can be included in a rule.

Step 3 - Set rule triggers.

Rule triggers enable you to define when a rule is launched (triggered). iLogic provides a list of event triggers to which the established rules are assigned.

- The list of triggers varies slightly depending on whether a part, assembly, or drawing is active.

- Each trigger provides you with standard functions that are commonly used, such as before a document save, when a document is closed, or when part geometry is changed.

- An iTrigger can also be used to trigger rules by adding a user parameter to the document that in turn launches any rules that it contains.

Step 4 - Create and edit rules, as necessary.

Continue to add or edit rules, as required.

- A complete list of rules can be reviewed in the iLogic browser.

- The order in which rules are listed in the iLogic browser can affect the resulting geometry. Drag and drop rules in the iLogic browser to capture the model's true design intent.

Practice 1a

Working with a Logical Model

Practice Objectives

- Review parameters in a part and assembly model.
- Review iLogic rules that have been created in part and assembly models.
- Launch and edit an iLogic form to modify the key parameters used to configure an assembly and its components.

In this practice, you will open an Autodesk Inventor assembly in which iLogic rules and forms have been created in the top-level assembly model. A component in the assembly also has iLogic-rules. Using a custom iLogic form, you will make changes to the model by selecting and entering new parameter values. The intent of the practice is to show how multiple design configurations can easily be created after iLogic rules have been incorporated into a model.

Task 1 - Open a part model and review its parameters.

1. If the Autodesk Inventor software is not open, select **Start> All Programs>Autodesk>Autodesk Inventor 2017> Autodesk Inventor 2017** or double-click on the **Autodesk Inventor 2017** icon on the desktop.

2. In the *Get Started* tab>Launch panel, click 📁 (Projects) to open the Projects dialog box.

This project file is used for the entire training material.

3. Click **Browse**, browse to *C:\Autodesk Inventor 2017 iLogic Practice Files* (or the directory of the installation files if you changed the default directory), and select **iLogic.ipj**. Click **Open**. The Projects dialog box updates and a checkmark displays next to the new project name, indicating that it is the active project. The project file tells the Autodesk Inventor software where your files are stored.

4. Click **Done**.

5. Open **Configured_Clip.ipt** in the *Overview* folder. The model displays as shown in Figure 1–6.

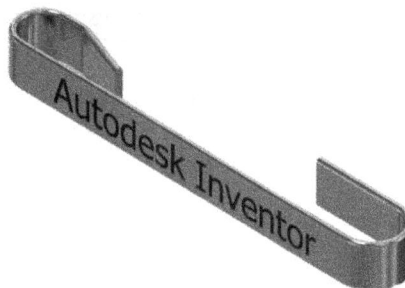

Figure 1–6

6. In the *Manage* tab>Parameters panel, click f_x (Parameters) to open the Parameters dialog box. Expand the **User Parameters** node, if required. Five key user parameters have been created in the model. These have been added to the model for use in the iLogic rules and contain the required configuration options. For example, the **Clip_Color** parameter enables you to select from five material types.

7. Expand the **Model Parameters** node, if required. The four key model parameters (**Length**, **Clip_Angle**, **Thickness**, and **Width**) are also used in the iLogic rules.

8. Close the Parameters dialog box.

Task 2 - Launch an iLogic Form to configure the model.

---

1. In the *Manage* tab>iLogic panel, click (iLogic browser) to toggle on the display of the iLogic browser. The browser displays embedded above or below the Model browser on the left side of the interface or might open as a floating browser. If floating, drag it to the bottom of the Model browser to dock it.

2. In the *Rules* tab in the iLogic browser, note that seven iLogic rules have been created in the model. Their descriptive names help to identify their purposes. For example, the **Clip_Color** rule controls the clip color based on the material type that is selected.

3. Double-click on the **Clip_Color** rule to open the Edit Rule dialog box. This dialog box is used to program all of the iLogic rules. Click **Cancel** to close the dialog box without making any changes.

4. Select the *Forms* tab. It contains any forms that have been created in the model.

5. Click **Clip Configuration** to open the Clip Configuration form, as shown in Figure 1–7.

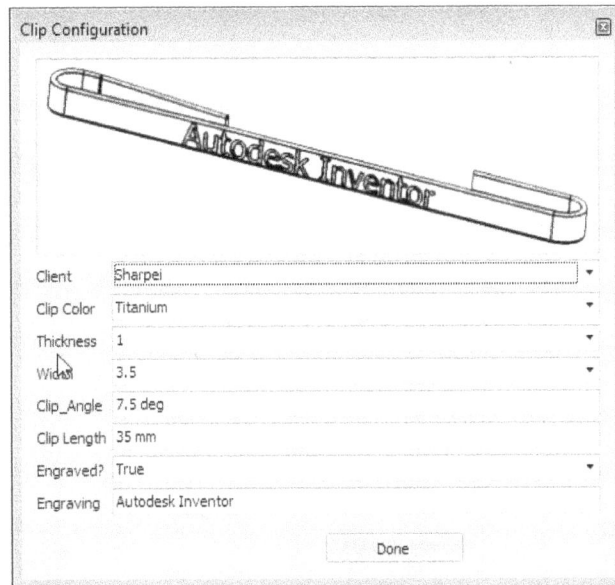

Figure 1–7

Each of the items in the form drives the following changes in the model:

- Assigns the client name for the project. This is a drop-down list of available client names.

- Defines the clip color. This is a drop-down list of available materials.

- Assigns the thickness and width of the clip. This is a drop-down list of the available sizes.

- Assigns the clip angle. This is a user-entry field. The permitted range of angular values is 7.5 to 20.

- Assigns the clip length. This is a user-entry field. The permitted range of values is 35 to 70mm.

- Assigns engraving to the surface of the clip. You can specify whether the engraving is to be included. If set to **True**, you can specify the engraving text. When engraving is included, a message displays details about the acceptable character length.

If outside of the permitted range, an assigned value is set and you are provided with feedback.

6. Change the model's configuration by selecting and entering values for each of the fields.

7. The model updates with the changes as you make selections in the drop-down lists, or when you press <Enter> after entering a value.

8. Click **Done** to close the form.

9. Close the model without saving.

Task 3 - Launch an iLogic form to configure the assembly.

1. Open **Mechanical Pencil_Complete.iam** in the *Overview* folder. The model displays as shown in Figure 1–8.

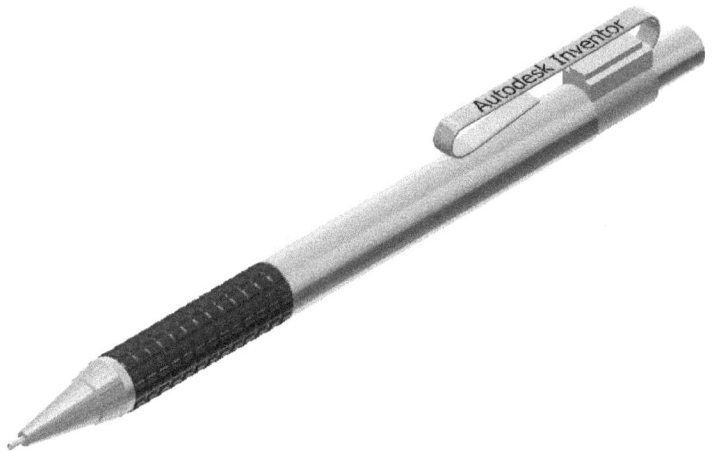

Figure 1–8

*To set the LOD representation, expand the **Representations** and **Level of Detail** nodes in the Model browser, and double-click on iLogic.*

2. The assembly should open in the iLogic LOD representation. If not, set the **iLogic** representation as the active Level of Detail. This representation is required for the Grip Style iLogic rule to run correctly. This is discussed in the Assembly functions content.

3. In the iLogic browser, select the *Rules* tab. The rules provided enable customization of the assembly.

4. Select the *Forms* tab. It contains any forms that have been created in the model.

5. Click **Assembly Configuration** to launch the Assembly Configuration form, as shown in Figure 1–9.

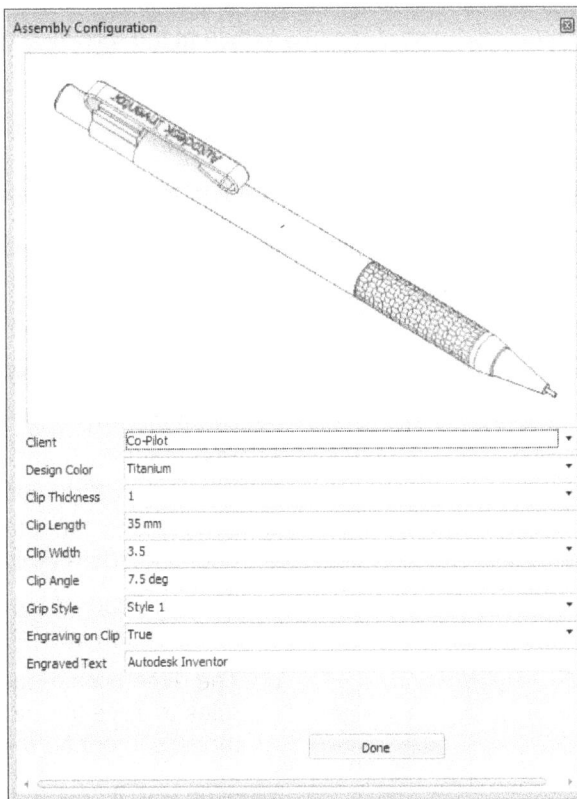

Figure 1–9

Each of the items in the form drive changes in the assembly and clip models. Many of the options are the same as in the clip model and the values entered here are pushed to the clip model to make changes to it. The only item that is unique to the assembly is the Grip Style. This controls how the grip geometry is represented in the assembly.

6. Change the assembly's configuration by selecting and entering values. The components display the changes as selections are made, or when you press <Enter> after entering a value.

7. Click **Done** to close the form.

8. Close the model without saving.

Chapter Review Questions

1. Which of the following design automation tools are available as part of the Autodesk Inventor software? (Select all that apply.)

   a. **Parameters**

   b. **iParts**

   c. **iAssemblies**

   d. **iLogic**

   e. **API Programming**

   f. **Autodesk Inventor Engineer-to-Order Series**

2. Which of the following best describes why it is important to ensure that the dimension scheme in a model captures the design intent before iLogic rules are added?

   a. Dimensions cannot be modified once iLogic rules have been incorporated into the model.

   b. Features cannot be modified once iLogic rules have been incorporated into the model.

   c. The dimension/model parameters are referenced in the rules to drive the model geometry. They must capture the model's intent for the rule referencing to work correctly.

3. Working with iLogic requires experience with API and VBA programming.

   a. True

   b. False

4. Which of the following can be accomplished using iLogic? (Select all that apply.)

   a. Perform multiple operations based on a single user input.

   b. Update material and iProperty information in the model.

   c. Provide user with customized feedback based on specified conditions.

   d. Drive iFeature, iPart, and iAssembly configurations.

   e. Interface directly with a predefined form to assign parameter values.

Understanding Parameters and Equations

The rules that drive iLogic are written based on model and user parameters. The model and user parameters tie the programmed logic to the model geometry so that the model reacts as programmed. Equations are also important because they can further establish mathematical relationships between dimensions and/or parameters.

Learning Objectives in this Chapter

- Change the dimension display type to identify dimension values, names, or equations.
- Create equations between dimensions to incorporate design intent into the model.
- Describe the difference between model parameters and user-defined parameters.
- Create user-defined parameters in a model.

2.1 Equations

Features and sketches generate dimensions in a model. Each dimension has a unique dimension name. The dimension name starts with the letter "d" followed by a unique number (e.g., d0 or d1).

Equations are relationships that can be defined between these dimensions, enabling you to control a dimension value based on a function of another dimension's value. Equations can also include user parameters. When one dimension is referencing another, the referenced dimension in the equation is considered the driving dimension. The model in Figure 2–1 shows a hole being located based on an equation.

*When **fx** displays as part of a dimension name, it indicates that the dimension contains an equation.*

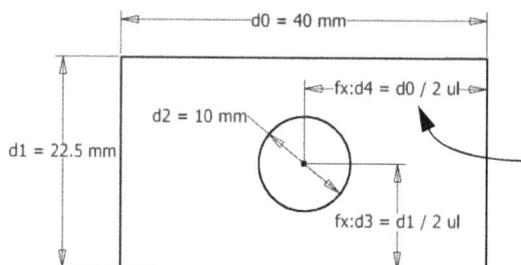

The distance from the center of the hole to the edges of the plate can be defined as half the overall dimension of the plate (i.e., d4=d0/2).

Figure 2–1

General Steps

Use the following general steps to add an equation:

1. Display the dimension symbols.
2. Create equations.
3. Flex the model.
4. Edit the equations, as required.

Step 1 - Display the dimension symbols.

To create equations you must use dimension names. To display dimension names, right-click in the graphics window, select **Dimension Display**, and click **Name**, as shown in Figure 2–2.

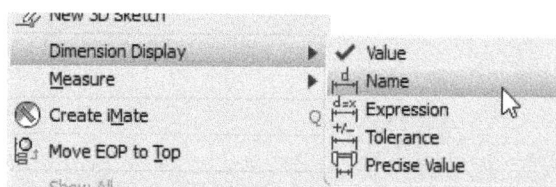

Figure 2–2

Additional styles are described as follows:

To control the display of dimensions in a Sketch, expand

(Dimension Display) in the Status Bar and select an option. The options are the same as those in the Part environment.

Value	Displays the numeric values of the dimensions.

Name	Displays the dimension names (e.g., d4 or d5).

Expression	Displays the dimension names with their numeric values or the equations that result in the numeric values (e.g., d2=d0/3 or d4=d5).

Tolerance	Displays the tolerances defined on any dimensions.
Precise value	Displays the precise numeric values of dimensions (not rounded). The **Precise value** option also displays the actual value of a dimension if a tolerance is applied. For example, if the dimension is 2.00 +/- 0.05 and is evaluated at the maximum tolerance, the value is 2.0500000000.

Step 2 - Create equations.

Equations can be created in a sketch or between features. They can be created using any one of the following techniques:

- When editing a sketch or feature dimension you can enter an equation directly in the Edit Dimension dialog box, as shown in the two examples in Figure 2–3. Alternatively, you can also select a dimension to add its dimension name to the *Edit Dimension* field for use in an equation.

*If **ul** displays next to a value, it indicates that the value is unitless.*

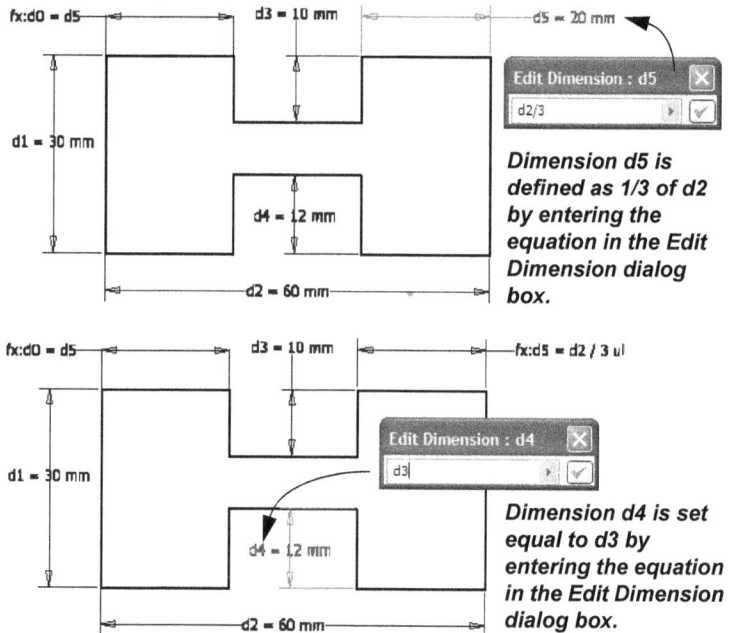

fx:d0 = d5   d3 = 10 mm   d5 = 20 mm

d1 = 30 mm

d4 = 12 mm

d2 = 60 mm

Edit Dimension : d5

d2/3

Dimension d5 is defined as 1/3 of d2 by entering the equation in the Edit Dimension dialog box.

fx:d0 = d5   d3 = 10 mm   fx:d5 = d2 / 3 ul

d1 = 30 mm

d4 = 12 mm

d2 = 60 mm

Edit Dimension : d4

d3

Dimension d4 is set equal to d3 by entering the equation in the Edit Dimension dialog box.

Figure 2–3

- During feature creation, you can enter an equation in the entry fields, as shown for the diameter and linear dimension in Figure 2–4.

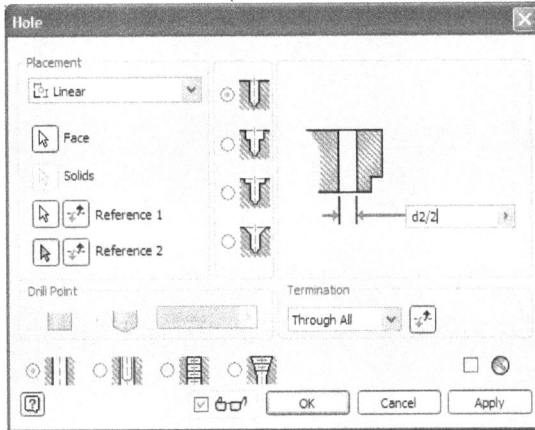

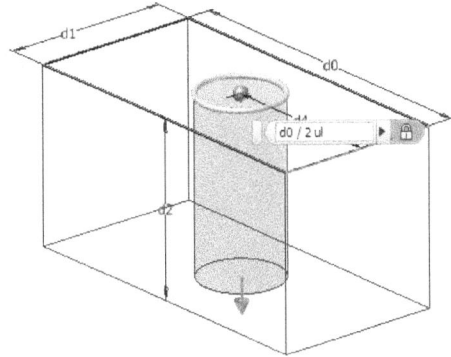

Figure 2–4

Model parameters can also be renamed in the Parameters dialog box to help recognize what the parameter is controlling.

• Equations can be entered after sketch or feature creation using the Parameters dialog box. This dialog box enables you to review the complete list of all of the parameters (dimension names) and existing equations in the model. To open the Parameters dialog box, in the *Manage* tab>

Parameters panel, click f_x (Parameters), or in the Quick

Access Toolbar, click $\boxed{f_x}$ (Parameters). Equations are entered in the *Equation* column, as shown in Figure 2–5.

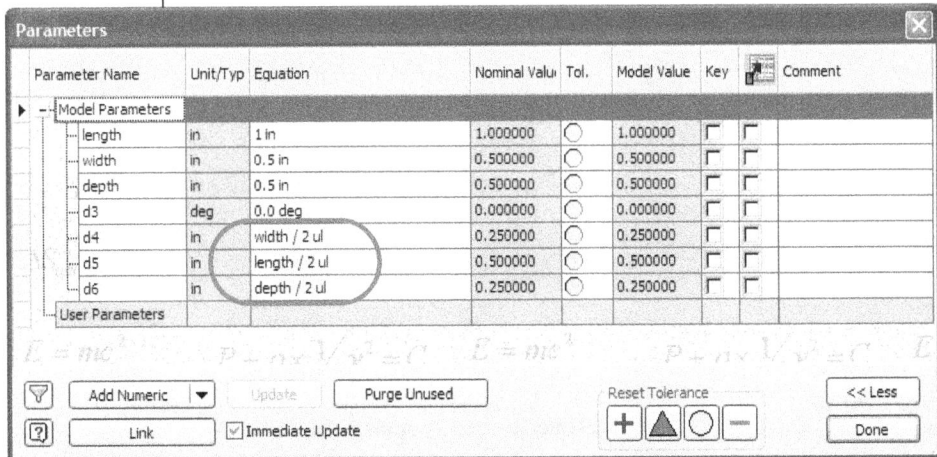

Parameter Name	Unit/Typ	Equation	Nominal Valu	Tol.	Model Value	Key		Comment
▶ – –Model Parameters								
┈ length	in	1 in	1.000000	○	1.000000	▯	▯	
┈ width	in	0.5 in	0.500000	○	0.500000	▯	▯	
┈ depth	in	0.5 in	0.500000	○	0.500000	▯	▯	
┈ d3	deg	0.0 deg	0.000000	○	0.000000	▯	▯	
┈ d4	in	width / 2 ul	0.250000	○	0.250000	▯	▯	
┈ d5	in	length / 2 ul	0.500000	○	0.500000	▯	▯	
└┈ d6	in	depth / 2 ul	0.250000	○	0.250000	▯	▯	
┈ User Parameters								

Figure 2–5

Equations can be written using any of the following operators or functions to capture the intent of the equation.

Mathematical Operators

The following operators can be used in equations:

+	Addition
-	Subtraction
/	Division
*	Multiplication
^	Exponentiation
()	Expression delimiter

Functions

The following functions can be used in equations:

sin( )	tanh( )	sinh( )
cos( )	sqrt( )	cosh( )
tan( )	log( )	ceil( ) converts arbitrary real numbers to close integers. The ceil function of a real number x, ceil(x) returns the next highest integer (e.g., ceil(3.2) =4).
asin( )	ln( )	
acos( )	exp( )	floor( ) converts arbitrary real numbers to close integers. The floor function of a real number x, floor(x) returns the next smallest integer (e.g., floor(3.8) =3).
atan( )	abs( )	

Units

The software can resolve equations that use some operators (+, -, *, /). However, for operators such as exponents, units should be assigned (e.g., 3in^2ul)

Units are assigned to each value, if you do not assign them first. Some of the symbols used for units of distance are as follows:

Unitless	ul
Inches	in, inch, or "
Feet	ft, foot
Meter	m, meter
Centimeter	cm
Millimeter	mm

Step 3 - Flex the model.

Once you finish adding an equation, test the model to verify that the equation captures the required design intent. This is called flexing the model and should involve editing the driving dimension values to verify that the model changes as expected.

Step 4 - Edit the equations, as required.

If flexing the model did not result in the required behavior, you can make required changes to the equations. This can be done by editing a dimension or feature, or using the Parameters dialog box.

2.2 Parameters

As features are created in a model, dimensions are used to define the model's shape. The name and value of the dimension are considered a model parameter. Model parameters are listed in the Parameters dialog box.

The Parameters dialog box also enables you to create user-defined parameters, which can be used in equations to help you control the model. Once created, user-defined parameters are listed under the **User Parameters** node in the Parameters dialog box.

To open the Parameters dialog box, use one of the following methods:

- In the *Manage* tab, on the Parameters panel, click f_x (Parameters).

- In the Quick Access Toolbar, select f_x (Parameters).

Model Parameters

Model parameters are the dimensions that are automatically assigned as you add sketch dimensions and features to the model. They use default dimension names (e.g., d0, d1, d2, etc.), and are listed at the top of the Parameters dialog box, as shown in Figure 2–6.

The sketch and feature dimension are listed in the Model Parameters node.

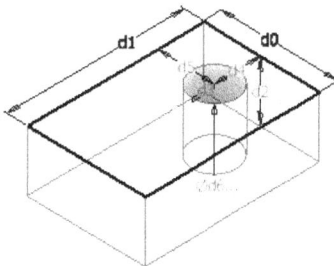

Parameter Name	Unit/Typ	Equation	Nominal Value	Tol.	Model Value	Key	E	Comment
Model Parameters								
d0	in	10 in	10.000000	○	10.000000	▢	▢	
d1	in	15 in	15.000000	○	15.000000	▢	▢	
d2	in	5 in	5.000000	○	5.000000	▢	▢	
d3	deg	0.0 deg	0.000000	○	0.000000	▢	▢	
d4	in	4 in	4.000000	○	4.000000	▢	▢	
d5	in	5 in	5.000000	○	5.000000	▢	▢	
d6	in	4 in	4.000000	○	4.000000	▢	▢	
User Parameters								

Figure 2–6

Renaming model parameters enables you to easily identify the parameter in the context of the entire model. To rename a model parameter enter a new name in the *Parameter Name* column of the Parameters dialog box, as shown in Figure 2–7.
Alternatively, you can enter a new name in the entry fields when creating or modifying dimensions associated with the feature you want to rename.

Model parameter names can be assigned when creating/editing the model or in the dialog box.

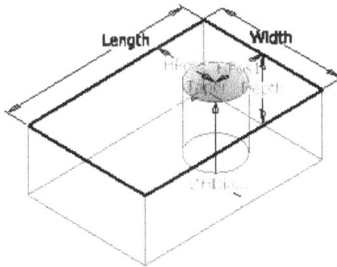

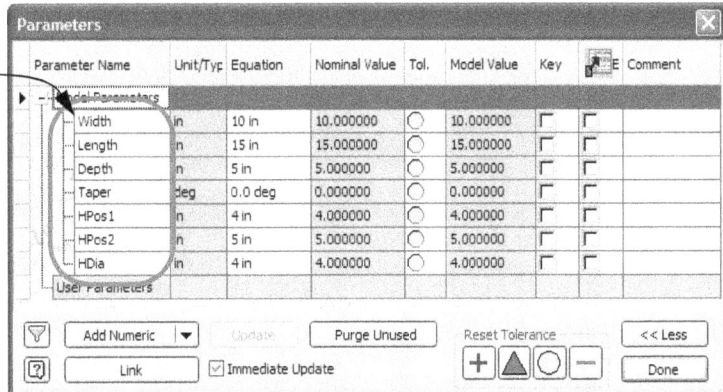

Figure 2–7

User
Parameters

User parameters can be added to further capture the model's design intent for use in equations or to add information to the model. User parameters can be any one of three types (i.e., Numeric, Text, or True/False) and are created in the Parameters dialog box.

To create a user parameter, select the option that is associated with the required parameter type at the bottom of the Parameters dialog box (i.e. **Add Numeric**, **Add Text**, **Add True/False**), as shown in Figure 2–8. Once added, the parameter displays in the **User Parameters** node at the bottom of the Parameters dialog box. Enter a name, unit (if applicable), and value for the new parameter.

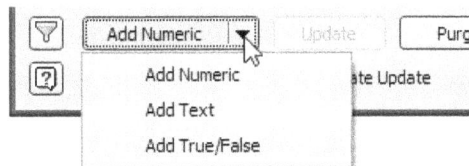

Figure 2–8

- Parameter names must be unique. They are case-sensitive and cannot begin with a numeric digit, nor can they contain spaces or mathematical operators.

- Some names are reserved by the software for specific operations and for mathematical use. If the name is unavailable for use you will be prompted that the parameter cannot be created.

Multi-value parameters can also be assigned for use with model parameters.

- Numeric and text parameters can be either a single-value or a multi-value parameter. To set a parameter as multi-value, right-click on any cell in its row and select **Make Multi-Value**. Using the Value List Editor (shown in Figure 2–9), multiple values can be added. Enter the possible values in the *Add New Items* area in the dialog box and click **Add** to include them in the multi-value list. Multi-value parameters display all of the available values in a drop-down list in the *Equation* row, as shown on the right of in Figure 2–9.

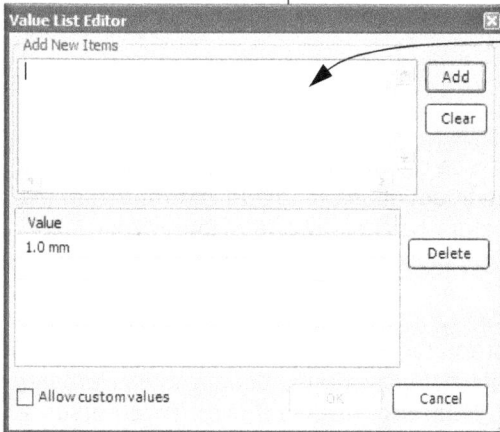

Enter multi-values in the Value List Edition so that they can be displayed in a list in the Parameters dialog box.

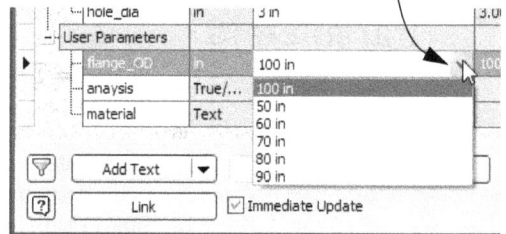

Figure 2–9

- Text parameters provide static text as their value.

- Similar to multi-value parameters, true/false parameters also display a drop-down list in the *Equation* row, enabling you to select the **True** or **False** value.

- To edit parameter values you can edit the model geometry (in the case of model parameters) or change the Equation cell value in the Parameters dialog box. Editing the value in the *Equation* column has the same effect as editing the dimension value directly on the model. User parameters can only be edited in the Parameters dialog box.

- When editing a dimension value using the *Edit Dimension* field, you can select from the list to efficiently create equations. Click ⬚ when editing and select **List Parameters** to open and directly select from the list of renamed parameters, as shown in Figure 2–10. Only renamed parameters are shown in this list.

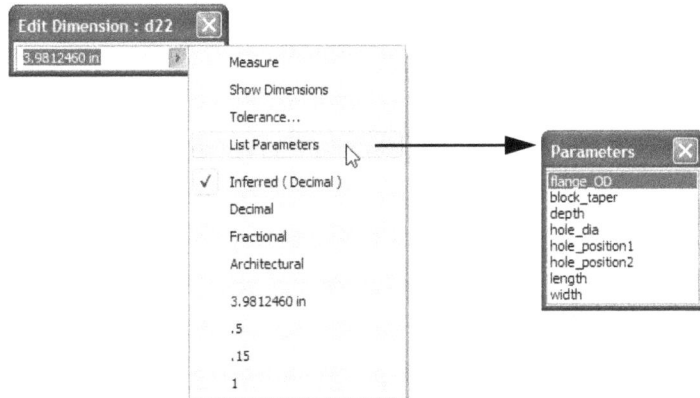

Figure 2–10

- Incorporating filters enables you to simplify the list of displayed parameters based on a selected filter type. To filter, click ⬚ (Filter) in the dialog box and select a filter type, as shown in Figure 2–11. The filtering options enable you to show only:

  - Key parameters (**Key**)
  - Non-key parameters (**Non-Key**)
  - Renamed parameters (**Renamed**)
  - Parameters driven by equations (**Equation**)

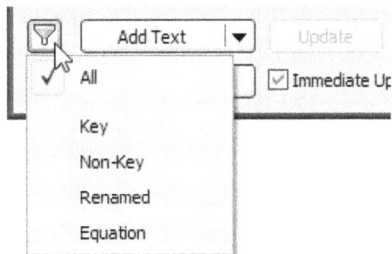

Figure 2–11

- To delete individual user parameters, right-click on the parameter name and select **Delete Parameter**. Alternatively, you can delete multiple unused parameters at once by clicking **Purge Unused** and selecting the parameters in the Purge Parameters dialog box.

- Some of the other options in the Parameters dialog box are as follows:

Tol column	Displays the evaluated tolerance setting: • ⊞ (Upper) • ◮ (Median) • ◯ (Nominal) • ⊟ (Lower) To change the setting, click an icon and select from the drop-down list.
Model Value column	Displays the actual calculated value for the parameter based on the tolerance setting.
Key column	Identifies existing model and user parameters and flags them as critical (key).
Export Parameter column	Adds the parameter to the custom properties for the model. Custom properties can be used in bill of materials and part lists.
Comment column	Enables you to add a comment about the parameter.
Link	Links a parameter to a spreadsheet or to another file (part, assembly, or sheet metal).
Update and **Immediate Update**	Enables you to specify whether or not the model updates immediately (**Immediate Update**) when a change is made. If the **Immediate Update** option is disabled, you must manually update after closing the Parameters dialog box.
Reset Tolerance	Changes the Tolerance setting for all of the parameters.
Less and **More**	Enables you to customize the display by displaying less or more of the dialog box.

Parameters and iLogic

Before an iLogic rule can be created, all of the required dimensions, parameters, and equations that the rule is going to reference must already exist in the model. The rules that drive iLogic are written based on dimensions, parameters, and equations. This information ties rules to the model geometry. Therefore, you should always consider the final design intent when you design the model.

- Verify that the feature dimensions capture the model's geometric intent and that they update, as expected, if changes are made.

- Additionally, verify that any user-defined parameters and equations that have been added define the required intent.

Practice 2a

Add Parameters

Practice Objectives

- Add equations to a model using the Parameters dialog box and feature dialog boxes.
- Create user-defined parameters for use in equations.

In this practice, you will modify dimensions in a part using model parameters and then add user-defined parameters. The completed model displays as shown in Figure 2–12.

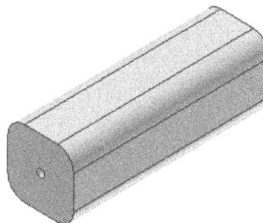

Figure 2–12

Task 1 - Open a part file.

1. Open **parameters.ipt**.

2. Right-click on **Extrusion1** and select **Show Dimensions**.

3. Click the left mouse button to clear the feature selection in the Model browser.

4. Right-click and select **Dimension Display>Expression**. The dimensions are displayed, as shown in Figure 2–13.

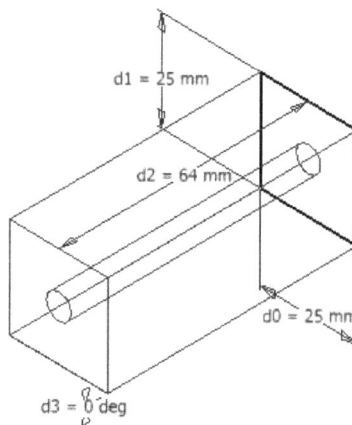

d1 = 25 mm
d2 = 64 mm
d0 = 25 mm
d3 = 0 deg

Figure 2–13

Task 2 - Change the model parameter names.

Alternatively, in the Manage tab> Parameters panel, click

f_x *(Parameters).*

1. In the Quick Access Toolbar, select $\boxed{f_x}$ (Parameters) to open the Parameters dialog box.

2. In the *Parameter Name* column, in the Parameters dialog box, change the following parameter names:

   • Rename *d0* as **side1**
   • Rename *d1* as **side2**
   • Rename *d2* as **length**
   • Rename *d4* as **dia**

3. Select the side2 *Equation* column and enter **side1**. The side2 dimension is now driven by the side1 value. The Parameters dialog box should be as shown in Figure 2–14.

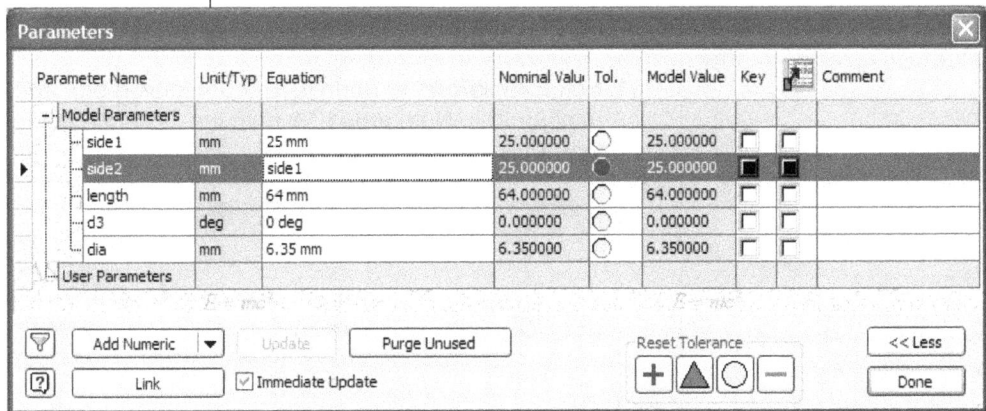

Figure 2–14

4. Click **Done** to close the Parameters dialog box.

5. Display the **Extrusion1** and **Hole1** dimensions. Note the parameter names that you assigned.

*To display dimensions for multiple features at the same time, select the features using <Shift>, right-click on any of the features in the Model browser, and select **Show Dimensions**.*

6. In the Quick Access Toolbar, select $\boxed{f_x}$ (Parameters) to open the Parameters dialog box again.

7. Select the length *Equation* column, enter **side1*3**, and press <Enter>. The entry in the *Model Value* column changes.

8. Click **Done** to close the Parameters dialog box.

To update the part, click

 (Local Update) in the Quick Access Toolbar.

9. Update the model, if required.

10. Display the **Extrusion1** dimensions. The model displays as shown in Figure 2–15.

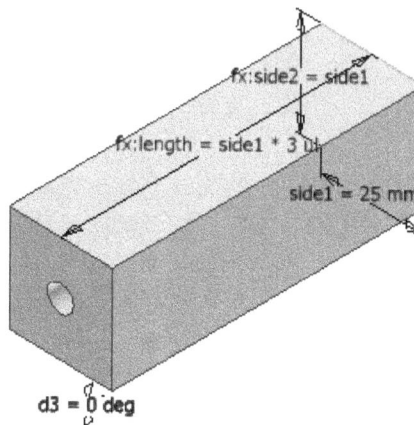

fx:side2 = side1

fx:length = side1 * 3 ul

side1 = 25 mm

d3 = 0 deg

Figure 2–15

11. Edit **Extrusion1** to open its Extrude dialog box and mini-toolbar. Note **side1 * 3 ul** in the *Distance* field, as shown in Figure 2–16.

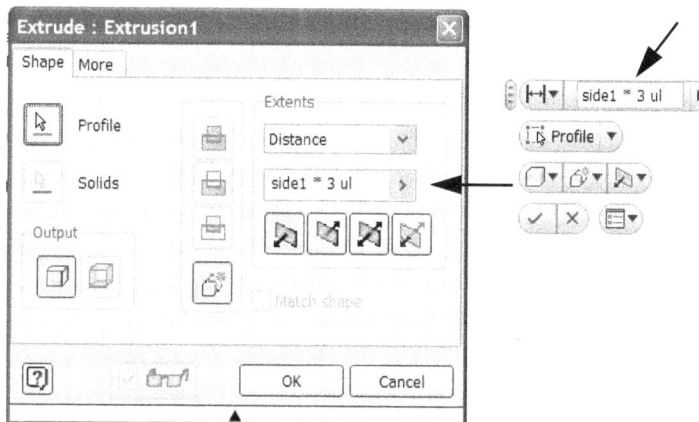

Figure 2–16

12. Close the Extrude dialog box.

Task 3 - Add a user-defined parameter.

1. In the Quick Access Toolbar, select f_x (Parameters) to open the Parameters dialog box.

2. Click **Add Numeric** in the Parameters dialog box to add a user-defined parameter to the model.

3. In the **User Parameters** node, in the *Parameter Name* column, type **size**.

4. Select the *Unit* column. The Unit Type dialog box opens.

5. Clear the *Unit Specification* field. Expand the Unitless branch and select **Unitless (ul)**.

6. Click **OK** to close the Unit Type dialog box.

7. In the **User Parameters** node, for the **size** parameter, set the following:

   • *Equation* column: **1 ul**
   • *Comment* column: **determine overall size**

8. Press <Enter>. The **size** parameter should display in the Parameters dialog box, as shown in Figure 2–17.

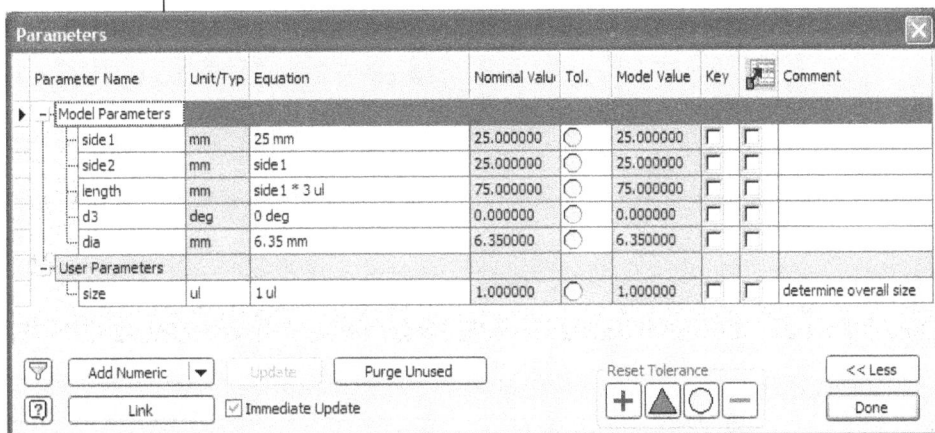

Parameter Name	Unit/Typ	Equation	Nominal Valu	Tol.	Model Value	Key		Comment
▶ – Model Parameters								
side1	mm	25 mm	25.000000	○	25.000000	☐	☐	
side2	mm	side1	25.000000	○	25.000000	☐	☐	
length	mm	side1 * 3 ul	75.000000	○	75.000000	☐	☐	
d3	deg	0 deg	0.000000	○	0.000000	☐	☐	
dia	mm	6.35 mm	6.350000	○	6.350000	☐	☐	
– User Parameters								
size	ul	1 ul	1.000000	○	1.000000	☐	☐	determine overall size

Add Numeric ▼   Update   Purge Unused   Reset Tolerance   << Less

Link   ☑ Immediate Update   + ▲ ○ −   Done

Figure 2–17

Task 4 - Add a second user-defined parameter.

1. Click **Add Numeric** to add a second user parameter.

2. In the second row of the **User Parameters** node, in the *Parameter Name* column, type **round_1**.

3. If units are not set to mm, select the *Unit* column, expand the Length branch, and select **millimeter (mm)**.

4. In the **User Parameters** node, for the **round_1** parameter, set the following:

   - *Equation* column: **3.175**
   - *Comment* column: **fillet radius**

5. Select the *Equation* cell associated with the **side1** parameter and type **25mm*size**. The Parameters dialog box should be as shown in Figure 2–18. The user-defined **size** parameter is going to drive the **side1** dimension parameter. Since the **size** parameter is unitless, you must multiply it by a distance to obtain a distance value.

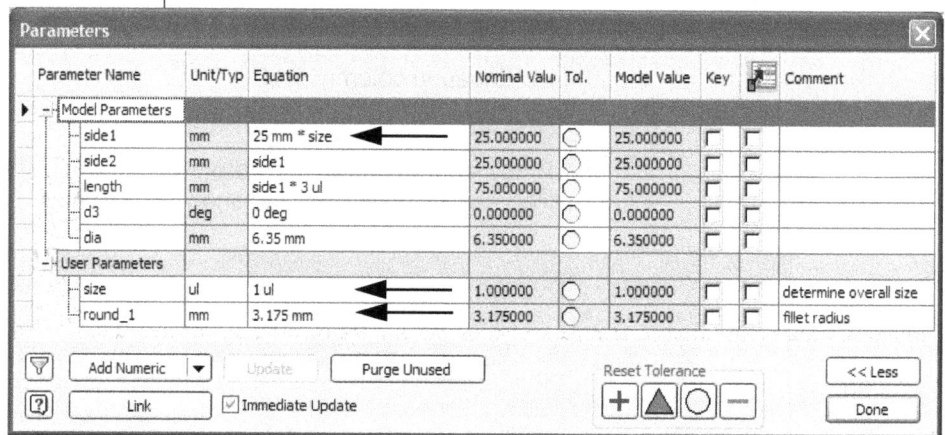

Parameter Name	Unit/Typ	Equation	Nominal Valu	Tol.	Model Value	Key		Comment
Model Parameters								
side1	mm	25 mm * size	25.000000		25.000000			
side2	mm	side1	25.000000		25.000000			
length	mm	side1 * 3 ul	75.000000		75.000000			
d3	deg	0 deg	0.000000		0.000000			
dia	mm	6.35 mm	6.350000		6.350000			
User Parameters								
size	ul	1 ul	1.000000		1.000000			determine overall size
round_1	mm	3.175 mm	3.175000		3.175000			fillet radius

Add Numeric ▼   Update   Purge Unused   Reset Tolerance   << Less

Link   ☑ Immediate Update   + ▲ ○ —   Done

Figure 2–18

6. Close the Parameters dialog box.

7. Display the **Extrusion1** dimensions. Note the parameter names and equations that you assigned.

Task 5 - Add fillets.

In this task, you will add fillets to the model and use the parameters you defined to modify the fillets.

1. Start the creation of the fillet.

2. Add fillets to the four long edges of the part.

3. Select the *Radius* field in the dialog box or in the mini-toolbar, and click ⬆. The fly-out menu displays as shown in Figure 2–19.

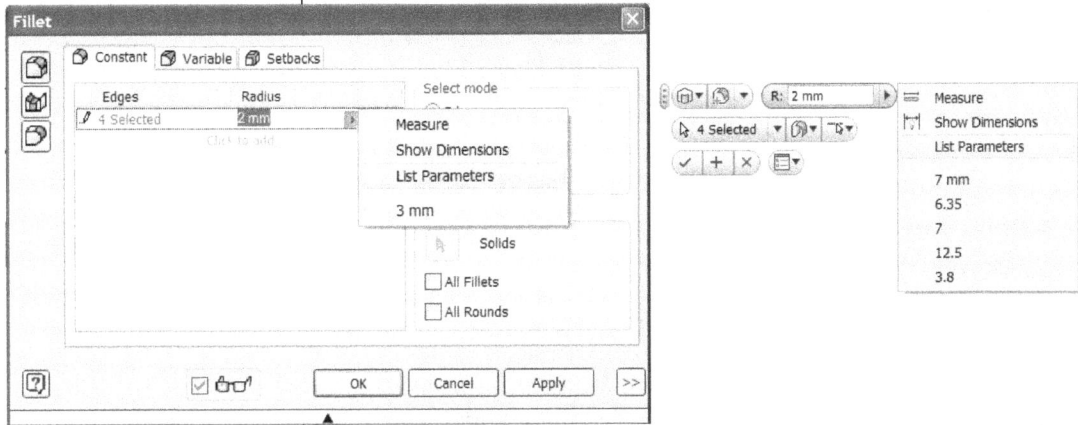

Figure 2–19

4. Select **List Parameters** and select **round_1** in the list that displays. The fillets radii are now controlled by the **round_1** parameter.

5. Complete the Fillet.

6. Change the equation of the **round_1** parameter to **12.5** in the Parameters dialog box.

7. Ensure that the **Immediate Update** option is selected in the Parameters dialog box. The model displays as shown on the left in Figure 2–20.

8. Change the equation of the **size** parameter to **2**. The updated model displays as shown on the right in Figure 2–20.

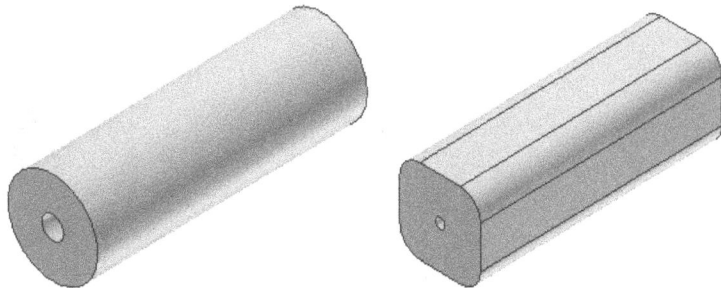

Figure 2–20

9. Save the model and close the window.

Practice 2b	# Building a Logical Part Model I

Practice Objectives

- Modify model parameter values by entering new values in the Parameters dialog box.
- Create a new user-defined parameter and establish a relationship between it and a model parameter so that it can be used to drive the geometry.
- Filter the list of parameters in the Parameters dialog box based on those that have been renamed from their default values.

In this practice you will prepare a model for use with iLogic functionality. The preparation will focus on establishing the parameters and equations that are required to meet the following design criteria:

- Depending on the flange's number of ribs and outer diameter, the rib geometry changes automatically. The outer diameter is to be selected from a list of six options ranging from 50 mm to 100 mm. The number of ribs can be entered manually.

- All rounds can be toggled on or off with a parameter setting.

- Model iProperties change depending on the material selection.

- A user-defined entry with a specific range for the height of the center hub can be used.

The initial model and some final variations that can be created after iLogic rules have been added are shown in Figure 2–21.

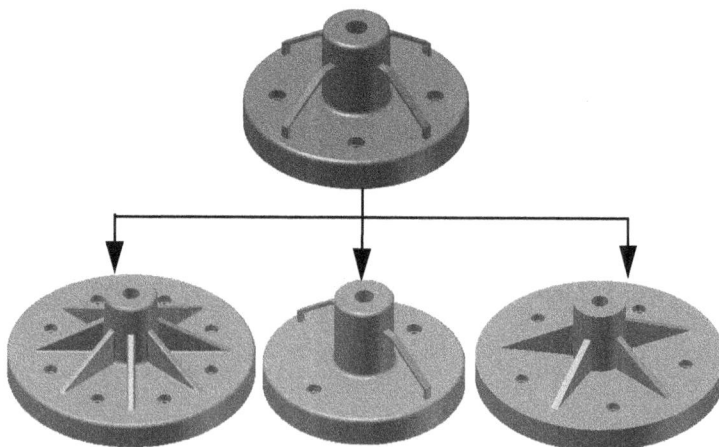

Figure 2–21

Task 1 - Open a part file.

1. Open **iLogic_flange.prt**.

2. In the Model browser, expand the **Patterned Ribs and Holes** feature. It is a pattern of two rib features, a hole, and any fillet that exists on the ribs. The Rib2 feature and its fillets are currently suppressed. Both ribs were created in the model as varying design options. The two rib configurations are shown in Figure 2–22.

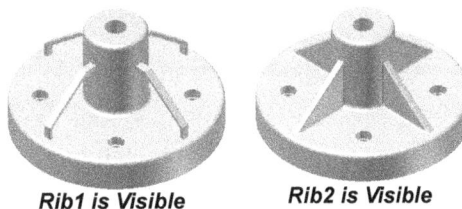

Rib1 is Visible   **Rib2 is Visible**

Figure 2–22

Task 2 - Review and add new user parameters to the model.

In this task, you will work in the Parameters dialog box to prepare the model for use with iLogic.

1. In the *Manage* tab>Parameters panel, click f_x (Parameters). The Parameters dialog box opens as shown in Figure 2–23.

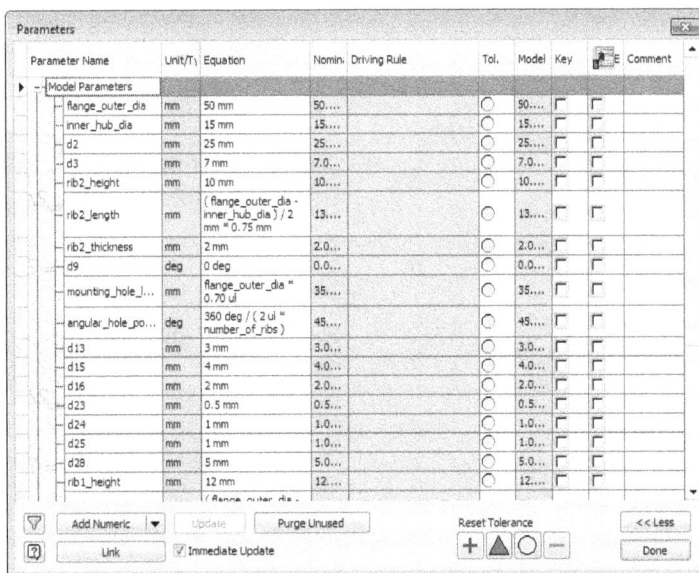

Parameter Name	Unit/T\	Equation	Nomin:	Driving Rule	Tol.	Model	Key		Comment
▶ — Model Parameters									
flange_outer_dia	mm	50 mm	50....		○	50....	☐	☐	
inner_hub_dia	mm	15 mm	15,...		○	15....	☐	☐	
d2	mm	25 mm	25,...		○	25....	☐	☐	
d3	mm	7 mm	7.0...		○	7.0...	☐	☐	
rib2_height	mm	10 mm	10....		○	10....	☐	☐	
rib2_length	mm	(flange_outer_dia - inner_hub_dia) / 2 mm * 0.75 mm	13,...		○	13....	☐	☐	
rib2_thickness	mm	2 mm	2.0...		○	2.0...	☐	☐	
d9	deg	0 deg	0.0...		○	0.0...	☐	☐	
mounting_hole_l...	mm	flange_outer_dia * 0.70 ul	35,...		○	35....	☐	☐	
angular_hole_po...	deg	360 deg / (2 ul * number_of_ribs)	45,...		○	45....	☐	☐	
d13	mm	3 mm	3.0...		○	3.0...	☐	☐	
d15	mm	4 mm	4.0...		○	4.0...	☐	☐	
d16	mm	2 mm	2.0...		○	2.0...	☐	☐	
d23	mm	0.5 mm	0.5...		○	0.5...	☐	☐	
d24	mm	1 mm	1.0...		○	1.0...	☐	☐	
d25	mm	1 mm	1.0...		○	1.0...	☐	☐	
d28	mm	5 mm	5.0...		○	5.0...	☐	☐	
rib1_height	mm	12 mm	12....		○	12....	☐	☐	
		(flange_outer_dia -							

Add Numeric ▼   Update   Purge Unused   Reset Tolerance   << Less

Link   ☑ Immediate Update   + ▲ ○   Done

Figure 2–23

2. Scroll through the list of parameters in the **Model Parameters** node. Note that a number of equations have already been set in the model to capture some of the design intent. These equations verify the following:

   - The equations for **rib2_length** and **rib1_length** are similar and ensure that the ribs extend a percentage of the difference between the inner and outer radius values.
   - The equation for **mounting_hole_location** positions the mounting holes at 70% of the overall **flange_outer_dia**.
   - The **angular_hole_position** sets the holes so that they are evenly spaced between the ribs, regardless of the number of ribs that are included in the model.

3. At the bottom of the Parameters dialog box, click ▽ and select **Renamed**. The parameters list is shortened and only displays the parameters that have been renamed. Filtering is useful in large models where the list of parameters can be long.

4. Currently, the model only contains four ribs. For the **number_of_ribs** parameter, select the *Equation* column, and type **6**. Note that the number of ribs, number of holes, and spacing of the holes update automatically in the model. This is because of equations that were previously assigned.

5. Return the **number_of_ribs** parameter value to **4**.

Model parameters are automatically generated as you create geometry. User parameters can be created as required, to capture the design intent. In the following steps, you will create a user parameter to provide a multi-value list for the flange's outer diameter.

6. Click **Add Numeric** to add a numeric user parameter in the **User Parameters** node in the dialog box.

7. For the name of the parameter, enter **flange_OD** and press <Enter>.

8. Right-click in any of the column cells for the **flange_OD** parameter and select **Make Multi-Value**. The Value List Editor displays.

9. In the lower area in the dialog box, select **1.0mm** and click **Delete** to remove it from the list.

10. Enter the values in the top area, as shown in Figure 2–24.

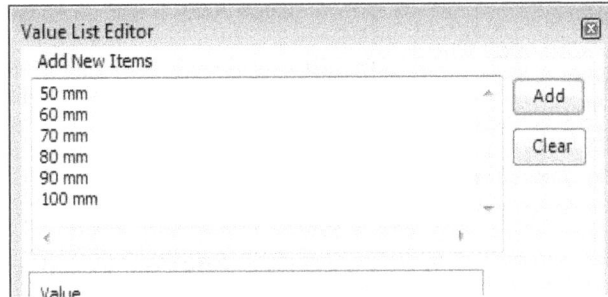

Value List Editor

Add New Items

50 mm		Add
60 mm		
70 mm		Clear
80 mm		
90 mm		
100 mm		

Value

Figure 2–24

11. Click **Add**. The values are all added to the lower area.

12. Click **OK**.

13. For the **flange_OD** parameter, select the *Equation* column. A drop-down list displays the six possible outer diameter values for the flange.

14. Select the **70 mm** value. Note that nothing changes in the model. This is because the **flange_OD** parameter is not tied to a dimension in the model.

*Alternatively, you could have established a multi-value list of values for the **flange_outer_dia** parameter.*

15. For the **flange_outer_dia** parameter, select the *Equation* column cell and type **flange_OD** as the equation. This establishes a relationship between the two parameters. Note that the model now updates to a 70 mm outer diameter, as shown in Figure 2–25. Note that the bolt circle of holes and size of the ribs also update because of equations that were assigned previously.

Figure 2–25

*With the **Immediate Update** option enabled, the model is recalculated each time a parameter change is made. To make multiple changes before updating, disable this option.*

16. Set the parameter value for **flange_OD** to **50 mm**.

17. Click **Add True/False** to add a user parameter.

18. For the name of the parameter, type **analysis**. This is a boolean parameter. It will be used in a future rule that will prepare the model geometry for analysis by suppressing all of the fillet features.

19. Set the parameter value for **analysis** to **False**.

20. Click **Add Text** to add a text parameter.

21. For the name of the parameter, type **material**.

22. Right-click in any of the cells for the **material** parameter and select **Make Multi-Value**. The Value List Editor displays.

23. In the top area in the dialog box, in separate lines, enter **Mild Steel** and **Stainless Steel**. Click **Add**. The values are added to the lower area in the dialog box.

24. Click **OK**. This list of material types will be used in a future rule to update the model's iProperties based on the material selection.

25. Set the **material** parameter value to **Mild Steel**.

26. Return the filter setting back to **All**.

27. Close the Parameters dialog box.

Task 3 - Add a new user parameter to the model.

In this task, you will modify the dimension scheme of the model to create a parameter that can be used in a rule that enables users to enter a height for the center hub. The current dimension scheme is not appropriate for the design intent.

1. Edit the sketch for the Base Feature Revolve feature.

2. In the Sketch, delete **d2**. This value controls the entire depth of the model, not just the height of the center hub.

3. Add a new dimension (as shown in Figure 2–26) and type **hub_height = 18** as its value. This assigns the parameter name for the dimension as **hub_height** so that you do not need to make the change in the Parameters dialog box.

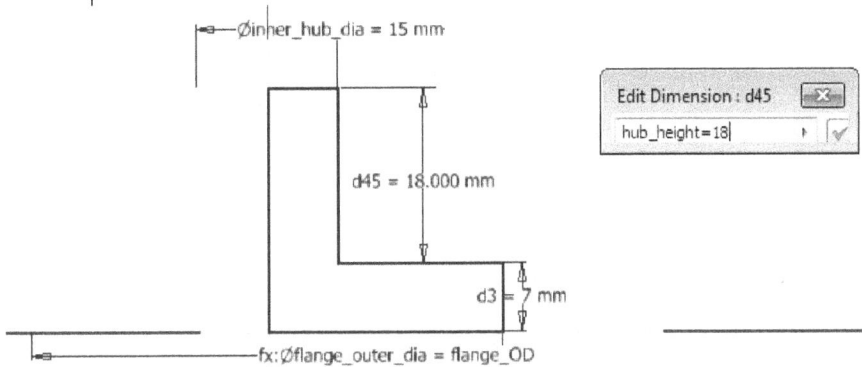

Figure 2–26

4. Finish the sketch.

5. Open the Parameters dialog box and review the renamed model parameter. Close the Parameters dialog box.

6. Save the part.

Chapter Review Questions

1. You can only add an equation to a model in the Parameters dialog box.

   a. True

   b. False

2. Which of the following statements regarding equations are true? (Select all that apply.)

   a. Dimensions and parameters can be used in an equation to drive a value.

   b. Equations can be manually entered in the Edit Dimension dialog box.

   c. Equations can be created by selecting dimensions directly from the model.

3. Which combination of equations centers the hole, if the length and width are equivalent, as shown in Figure 2–27? (Select all that apply.)

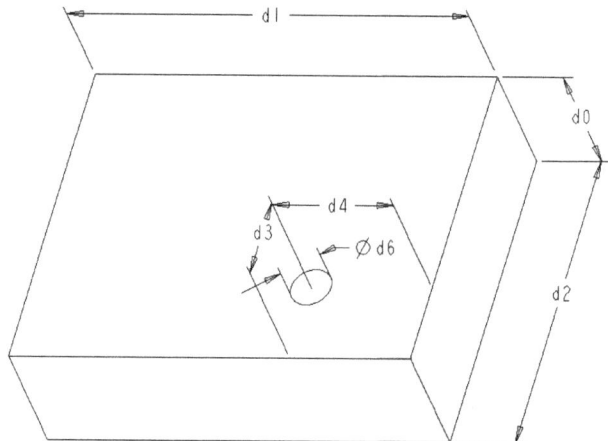

Figure 2–27

   a. d4 = d1 / 2
      d3 = d2 / 2

   b. d1 = d4 / 2
      d2 = d3 / 2

   c. d1 = d2
      d4 = d1 / 2
      d3 = d2 / 2

4. Which dimension display type is set to display the dimensions in the model as shown in Figure 2–28?

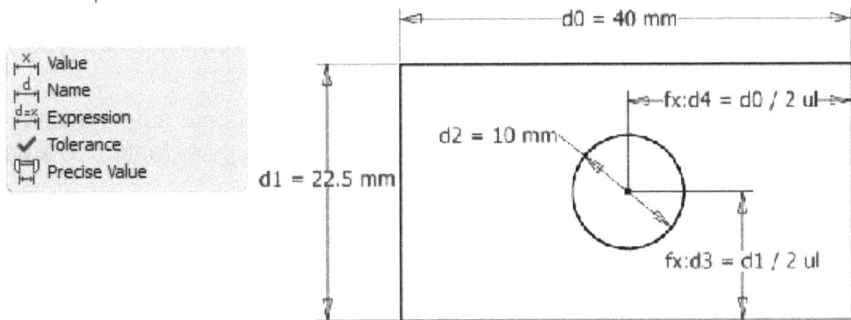

Figure 2–28

a. Value

b. Name

c. Expression

d. Tolerance

e. Precise Value

5. Which description best describes how the equation shown in Figure 2–29 affects the model?

$$d0 = 3 * d6$$
$$d3 = d4$$

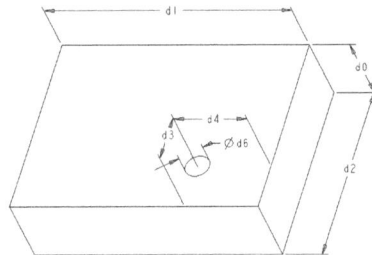

Figure 2–29

a. The depth of the base extrusion is equal to the diameter of the hole and the hole is centered on the base extrusion.

b. The depth of the base extrusion is equal to three times the diameter of the hole and the hole is centered on the base extrusion.

c. The depth of the base extrusion is equal to three times the diameter of the hole, and the values of the horizontal and vertical hole dimensions are equivalent.

d. The depth of the base extrusion is equal to three times the diameter of the hole, the values of the horizontal and vertical hole dimensions are equivalent, and d1 and d2 are equivalent.

6. Fill in the parameter type: Model or User-Defined, with the method in which it is created in a model.

  a. _____ parameters are created using the Parameters dialog box where you select the type, name, and enter values.

  b. _____ parameters are automatically assigned each time you add a dimension or feature to a part.

7. Which of the following best describe what **fx** is identifying in the sketch shown in Figure 2–30?

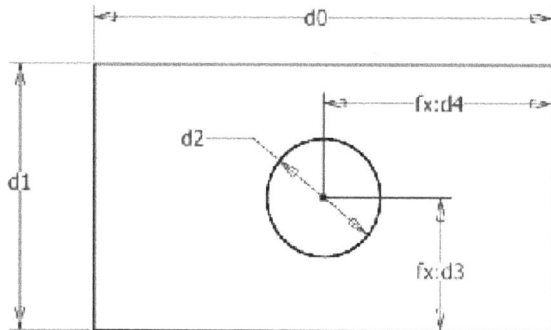

Figure 2–30

  a. d3 and d4 are reference dimensions.

  b. d3 and d4 are equal.

  c. d3 and d4 are generated based on a user-defined equation.

  d. d3 and d4 have a tolerance assigned.

8. Equations can be established during feature creation using the Feature Creation dialog box or mini-toolbar.

  a. True

  b. False

Command Summary

Button	Command	Location
f_x	Parameters	• **Ribbon:** *Manage* tab>Parameters panel • **Quick Access Toolbar**

Chapter

3

Getting Started with iLogic

A logical model, driven by iLogic rules, is modeled using normal modeling techniques and is then automated using the Edit Rule dialog box to create rules in the model. Using this dialog box, you can follow a general rule creation workflow to guide you through the process.

Learning Objectives in this Chapter

- Navigate to the iLogic panel in the interface to access iLogic commands.
- Control the display of the iLogic browser in the Autodesk® Inventor® interface.
- Describe the three areas in the Edit Rule dialog box and their role in assigning functions to create iLogic rules.
- Create a rule with conditional statements and functions that drive or read information from model geometry.

3.1 iLogic Interface

When learning how to use and create rules with Autodesk Inventor iLogic, you must be familiar with the interface components.

iLogic Panel

All of the iLogic commands are located in the *Manage* tab>iLogic panel, as shown in Figure 3–1. Additional options are available in the expanded panel by clicking ▼ in the Panel's heading.

The options available in the panel are discussed throughout this student guide.

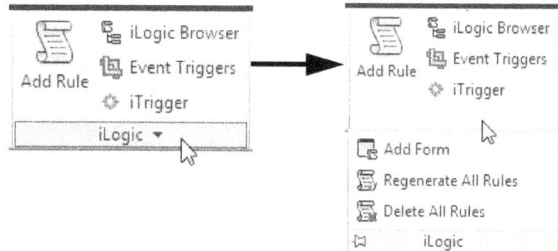

Figure 3–1

Edit Rule Dialog Box

iLogic programming is written with rules using the Edit Rule dialog box, as shown in Figure 3–2.

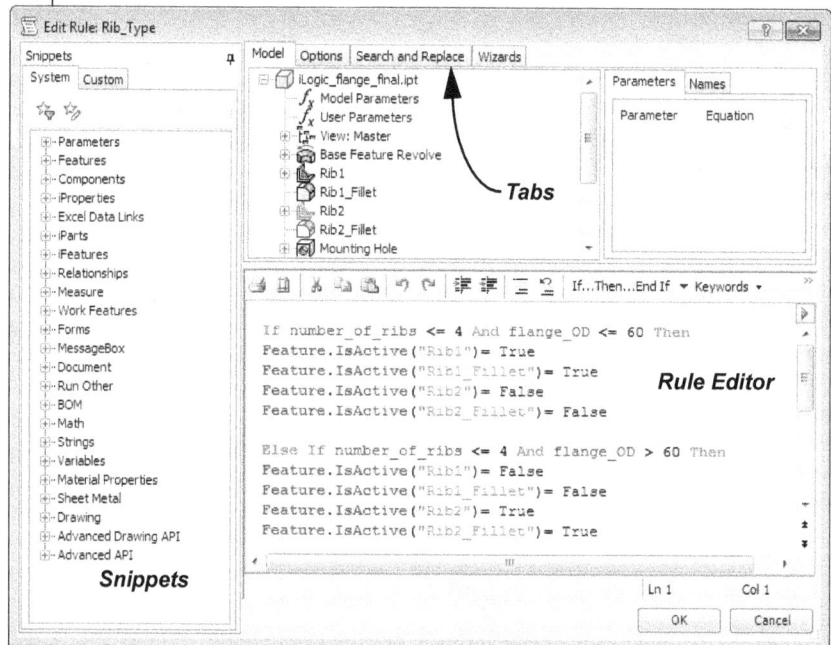

Figure 3–2

To create a rule you use a combination of all of the areas in the dialog box. In general, you use the Rule Editor to create the body of the rule. The *Snippets* and *Tabs* areas can be used to locate and help create the text in the rule.

Rule Editor

The *Rule Editor* area in the Edit Rule dialog box is the one in which all of the programming is completed.

- It is located in the bottom right in the Edit Rule dialog box.

- It contains standard and advanced editing tools that enable you to do such things as print, cut, copy, paste, undo, and redo.

The editing tools in the Rule Editor are shown in Figure 3–3.

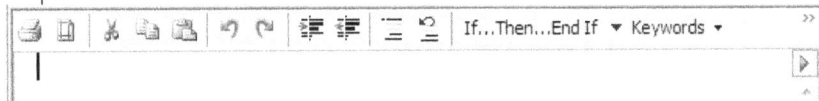

Figure 3–3

The editing tools in the Rule Editor are as follows:

Icon	Command	Description
	Print	Prints the text in the active rule.
	Page Setup	Customizes the page setup for printing the rule's text.
	Cut	Removes the selected rule text and places it on a clipboard for reuse in the current or a different rule.
	Copy	Copies the selected rule text and places it on a clipboard for reuse in the current or a different rule.
	Paste	Pastes the text that was previously placed on the clipboard.
	Increase/ Decrease Indent	Increases or decreases the spacing to the left of the current line in the Rule Editor. This can help when organizing code.
	Comment/ Uncomment lines	Adds or removes a comment tag in a selected line of text in the Rule Editor.

and enable you to organize code by indenting. Alternatively, or in conjunction with the tools, you can use the '[and '] lines of code to define a portion of code in an expandable node.

If...Then... End If		Provides a list of predefined conditional statements that can be added as text to a rule.
Keywords		Provides a list of predefined keywords that can be added as text to a rule. For example, Select Case, While, Return, etc.
Operators		Provides a list of predefined operators that can be added as text to a rule. For example, And, Or, >, <, etc.
⊚	Help	Displays Help that is available for the Edit Rule dialog box.

Rules are created as a combination of text elements, including conditional statements, parameters, and functions. Each of these elements of a rule are made easily identifiable through the use of colors. The rule shown in Figure 3–4 indicates how colors are used in a rule.

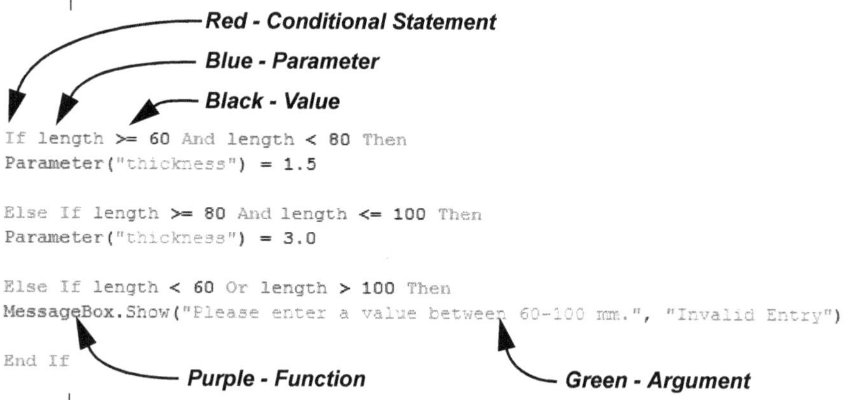

Figure 3–4

The colors used in a rule and the elements they help to identify are as follows:

Color	Description
Red	When conditional statements are added to a rule using options in the If...Then...End If drop-down list, the conditional statements keywords are displayed in red.
Brown	When conditional statements are added to a rule using options in the If...Then...End If drop-down list, locations where user entry is required are displayed in brown. The placeholder is indicated as **My_Expression**. For example, in the If-Then-End If statement (shown on the top of Figure 3–5), **My_Expression** is displayed in brown. The mistyped parameters in a rule also display in brown.

Black	The operator and value portions of a conditional statement or function are displayed as black. For example, in the If-Then-End If statement shown on the bottom of Figure 3–5, the operator and value for the **number_of_ribs** parameter is displayed in black.
Purple	The function element in a rule is displayed in purple.
Green	The argument in a function is displayed in green. This field must be edited to fully define the function. The color of the argument remains the same after editing. For example, in the If-Then-End If statement (shown on the top of Figure 3–5) "featurename" is displayed in green. Once Rib1 is included as the feature (that is affected by the function), it remains green, as shown at the bottom of Figure 3–5.

The rule shown at the top of Figure 3–5 is an initial rule statement using a default conditional statement, function, and argument. The rule shown at the bottom of Figure 3–5 has been edited to include the condition, argument, and values that connect the rule to the model.

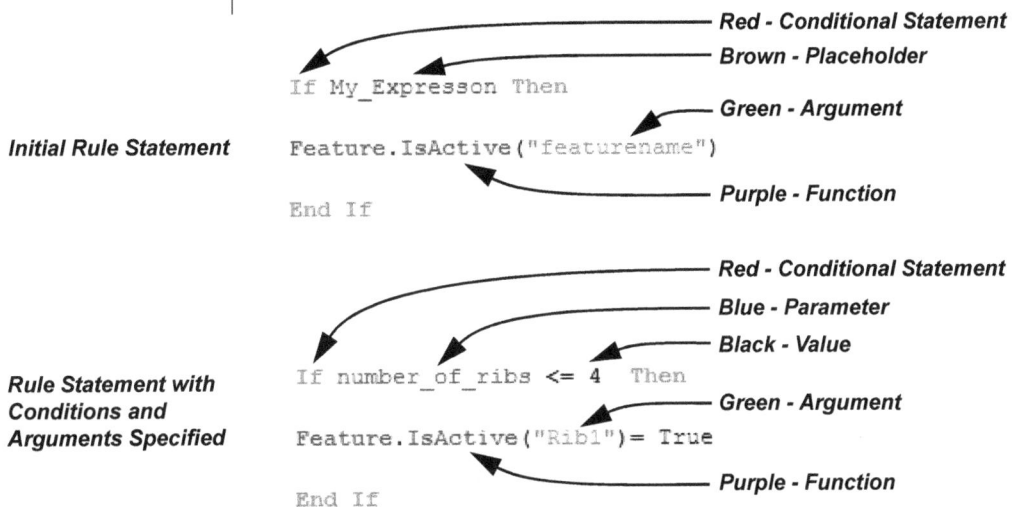

Initial Rule Statement

```
If My_Expresson Then

Feature.IsActive("featurename")

End If
```

- Red - Conditional Statement
- Brown - Placeholder
- Green - Argument
- Purple - Function

Rule Statement with Conditions and Arguments Specified

```
If number_of_ribs <= 4   Then

Feature.IsActive("Rib1")= True

End If
```

- Red - Conditional Statement
- Blue - Parameter
- Black - Value
- Green - Argument
- Purple - Function

Figure 3–5

The text in a rule can be entered in the following ways:

- Manually typing the text.

Reusing code helps eliminate typographical errors that can cause a rule to fail.

- Using the *Snippets* area in the *Model* tab, or the keywords or operators drop-down lists in the Rule Editor. This enables you to use saved functions, parameter names, and keywords in the rule.

Snippets

*This area is called Snippets because it provides the functions that constitute a portion or **snippet** of a rule.*

Every rule consists of functions that provide instructions on what is to be done in a model when a condition is met. The functions that can be used in an iLogic model are listed in the *Snippets* area in the Edit Rule dialog box, as shown in Figure 3–6.

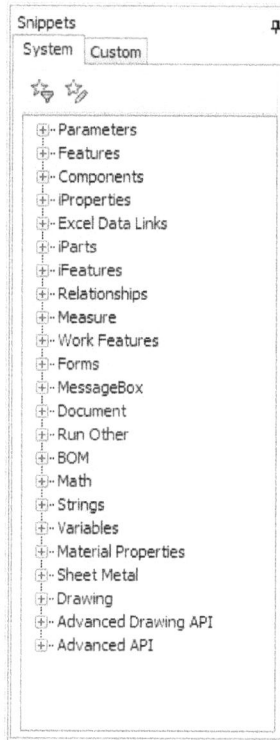

Figure 3–6

The *Snippets* area is divided into two tabs: *System* and *Custom*.

- The *System* tab provides a full list of the predefined functions. All of the functions are sub-divided into categories for organization.

- The *Custom* tab provides access to saved snippets.

You can move or hide the *Snippets* area using the following:

- Click (Pin) in the right corner in the *Snippets* area to toggle between auto-hide and docked.

- When auto-hidden, hover the cursor over the Snippets heading to expand it and access the functions.

- Undock the *Snippets* area by selecting its heading and dragging it outside the Edit Rule dialog box.

- Dock the *Snippets* area by selecting its heading and dragging it over the appropriate docking icon. (For example, drop it on

  [icon] to dock it on the left side of the Edit Rule dialog box.)

Tabs

The tabs in the top right area provide methods of adding content to a rule, general options for rule behavior, search tools, and additional predefined codes for a rule. Each of these tabs is discussed in depth throughout the student guide.

The *Model* tab (shown in Figure 3–7) is the most commonly used tab.

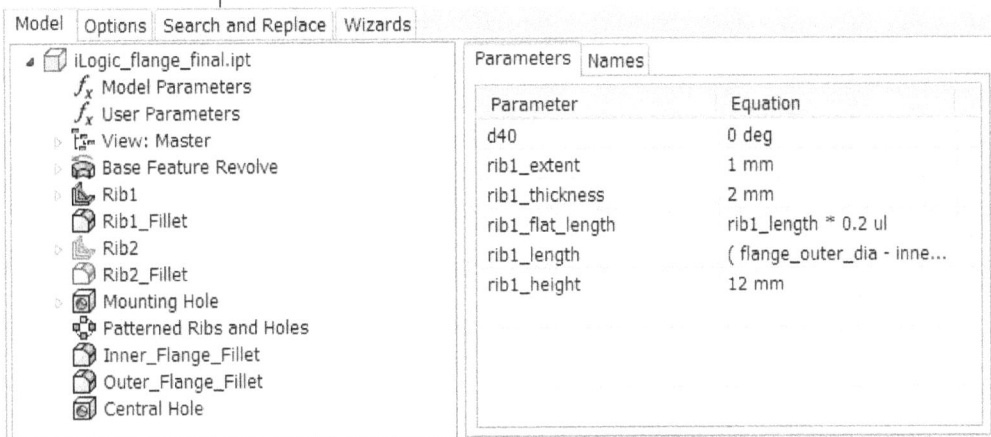

Parameter	Equation
d40	0 deg
rib1_extent	1 mm
rib1_thickness	2 mm
rib1_flat_length	rib1_length * 0.2 ul
rib1_length	(flange_outer_dia - inne...
rib1_height	12 mm

Figure 3–7

The *Model* tab is divided into left and right panes.

- The left pane is a copy of the Model browser. It lists all of the views and features, and provides access to any of the model and user parameters that exist in the model. To tie functions to specific geometry in the model, parameter and feature names are referenced inside most functions. This tab enables you to navigate through the left pane while displaying a selected item's parameters or feature names in the right pane. Once displayed, you can double-click on the parameter or feature name to pull the exact information into a rule.

- The syntax in a rule must be exact. Using the *Model* tab for parameter and feature name selection is recommended.

iLogic Browser

The iLogic browser is an important part of how you interact with existing rules and forms in a model. Similar to the Model browser, it is a browser that displays on the left side of the Autodesk Inventor interface and is vertically aligned with the Model browser. By default, the iLogic browser is not displayed. To toggle on the display of the iLogic browser, in the *Manage* tab>iLogic panel, click ⬚ (iLogic Browser). Figure 3–8 shows an iLogic browser for a model that already contains four rules.

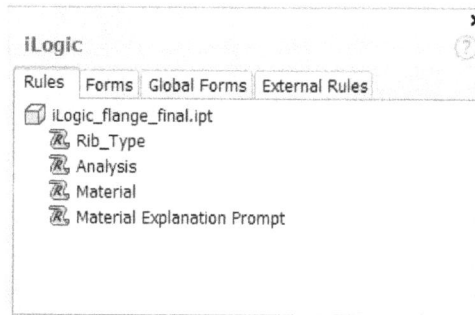

Figure 3–8

All tabs are displayed regardless of whether rules or forms exist.

In the iLogic browser, using the tabs along the top, you can do the following:

- Right-click on an existing rule to open it for editing, run or regenerate the rule, delete the rule, or suppress the rule.

- Right-click in white space in the browser to create a new rule, run, or regenerate all of the rules.

- Drag and drop rules in the list to ensure that the order of rule execution matches the required intent of the rules.

- Create new forms and edit existing forms in the current model or in external models.

- Access the list of rules in external models.

Once the iLogic browser displays, it remains active until it is explicitly removed from the display (between models and sessions). If a model or document is opened that does not support or contain iLogic rules or forms, the browser is displayed as empty. To remove the iLogic browser from the display, click **X** in the top right corner of the browser.

3.2 Function Overview

The Snippets area is called Snippets because it provides the functions that constitute a portion or snippet of a rule.

The basis of all of the rules are the instructions that are carried out in a program when a condition is met. The instructions that are programmed in the rule are called functions. The instructions in a function are what interacts with the model to read/write data or change the geometry. A substantial and varied list of functions are provided in the Autodesk Inventor software. These default snippets are listed in the *Snippets* area in the Edit Rule dialog box and sub-divided by category, as shown in Figure 3–9.

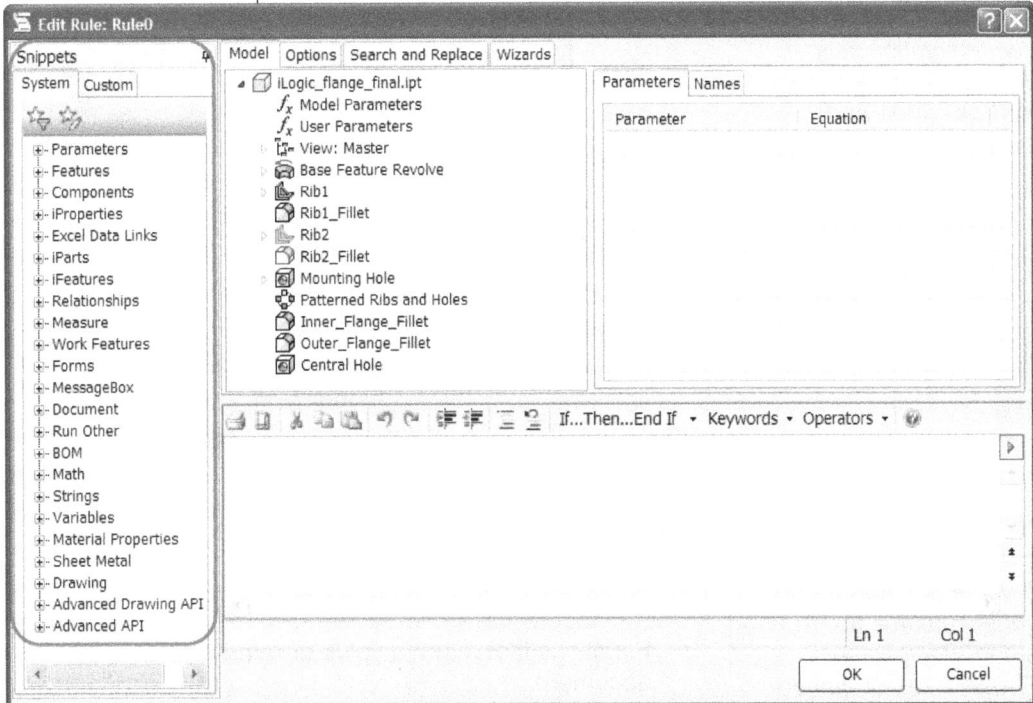

Figure 3–9

- All System snippets are stored in an .XML file. The name and location of the default .XML file that stores these snippets is *C:\Program Files\Autodesk\Inventor2017\Bin\en-US\iLogic Snippets.xml*. Editing this file is NOT recommended. Create a copy and edit that, if required.

Adding Snippets

Many functions use a common syntax for which the category and function name are included (**Category.FunctionName**). For other functions, the actual function might be more cryptic. Once a snippet category has been expanded, you can hover the cursor over any of the functions to display its syntax and a short help line. For example, in Figure 3–10 the cursor is hovered over the **Features>IsActive** function, indicating that it can be used to get or set activity of a feature (i.e., whether a feature is active or not) and its syntax is **Feature.IsActive ("featurename")**.

*Help provides full descriptions on all of the snippets. To locate this documentation, launch the Help system, search for **Functions** and select the topic.*

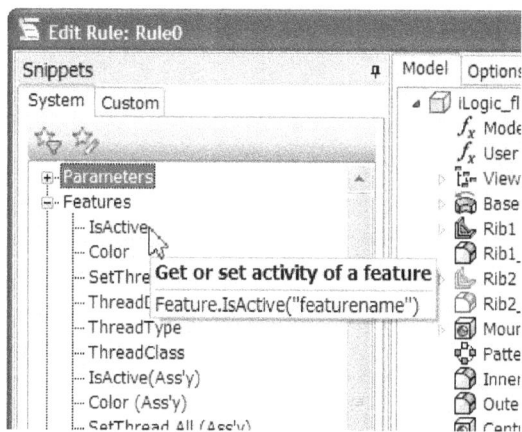

Figure 3–10

There are a number of different types of functions in the list of snippets. The different types of functions vary depending on whether they return a value, assign a value, or do neither. The function types are as follows:

A value or a returned value can be a text string, Boolean value (True/False), or number.

- Functions that assign a value write it to the document/model.

- Functions that return a value or retrieve information from a document.

- Functions that neither assign or return a value are called *Sub functions*. In general this facilitates user input and information display for other functions.

How To: Use a Function in a Rule

Functions in a rule can be copied and pasted in the rule or between rules as an alternative to creating it from scratch.

1. Enter the function into the rule.
   - Double-click the functions in the *Snippets* area.
   - Manually type the function. The syntax for functions must be exact. Therefore, selection is recommended.

2. Edit the arguments and values to fully define the function.

- Both arguments and values can be text strings, Boolean values (True/False), or numbers.
- In general arguments are included in a pair of double quotes (e.g., "length", "part1:1", "d12").

The default syntax and examples of each type of function are as follows. Note that the provided functions can be used to read or write values depending on requirements.

Function Type	Syntax	Examples
Write	Category.FunctionName(argument1, argument2, ...) = value	• Parameter (Dynamic) --> Parameter ("d0") = 1.2 • Features (IsActive) --> Feature.IsActive("featurename") = T/F • Color --> Feature.Color("featurename") = red • Material --> iProperties.Material = "Steel, Mild"
Read	returnValue = Category.FunctionName(argument1, argument2, ...)	• Parameter (Dynamic) --> parameter = Parameter ("d0") • iProperties.Value --> parameter = iProperties.Value ("part1:1", "Project", "Part Number") • Angle --> parameter = Measure.Angle("entityName1", "entityName2") • Area --> parameter = Measure.Area ("SketchName")
Sub	Category.FunctionName(argument1, argument2, ...)	• MessageBox (Show) --> MessageBox.Show ("To edit Material, select a value for the Material parameter in the Parameters dialog box.", "Material Change Notice") • Forms (ShowForm) --> iLogicForm.Show ("FormName")

In situations where functions are dependent on one another, verify the correct order in which the Functions are listed.

Favorite Snippets

Commonly used snippets can be assigned as favorites to help filter the list. To edit the favorites list, click ⭐ (Edit favorite snippets) and enable or disable the snippets so that only the favorites are enabled. Once defined, you can use ⭐ (Toggles the filtering of the system snippets to favorites only) to enable or disable the favorites only view. By default, the favorites list has a preset list of commonly used snippets.

Custom Snippets

The *Custom* tab provides advanced programming capabilities, where you can create and edit personal snippets to meet your design requirements. The *Custom* tab provides some default code in the *My Snippets* category, as shown in Figure 3–11.

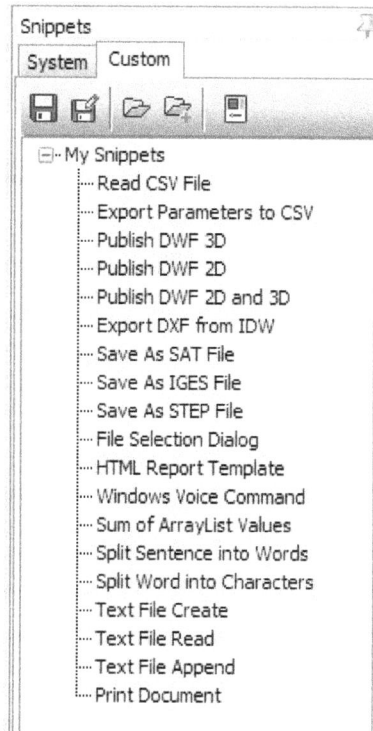

Figure 3–11

All snippets (System and Custom) are stored in an .XML file. The name and location of the default .XML file that stores the custom snippets is *C:\Program Files\Autodesk\Inventor 2015\Bin\en-US\iLogicUserSnippets.xml.* This file is active until another file is loaded. Any new custom snippets are saved to the active .XML file. The toolbar along the top of the *Custom* tab enables you to save new .XML files and load alternate .XML files. All of the snippets can be stored in one .XML file or you can create multiple files for use in different projects.

How To: Create a custom snippet

Custom snippets can include both conditional statements and functions.

1. In the Rule Editor, select the rows of text that you want to save as a custom snippet.
2. Right-click and select **Capture Snippet**, as shown in Figure 3–12.

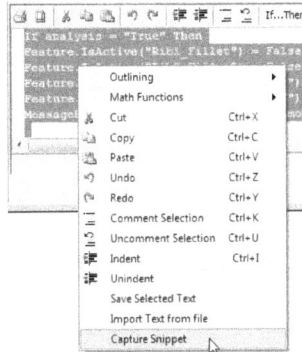

Figure 3–12

3. The Edit Snippet dialog box opens and the captured snippet is automatically copied into the Code text box. Enter the title of the snippet, as shown in Figure 3–13.

4. Select a Category in which to add the custom snippet. You can also enter a name to create a new category.

5. (Optional) Enter a text string in the *Tooltip* area, as shown in Figure 3–13. This text displays as the tooltip when you hover the cursor over the snippet. Alternatively, click **Use Code as Tooltip** to use the code as the tooltip. Once copied, you can edit to further customize the tooltip.

6. Click **OK** in the Edit Snippet dialog box to complete the custom snippet.

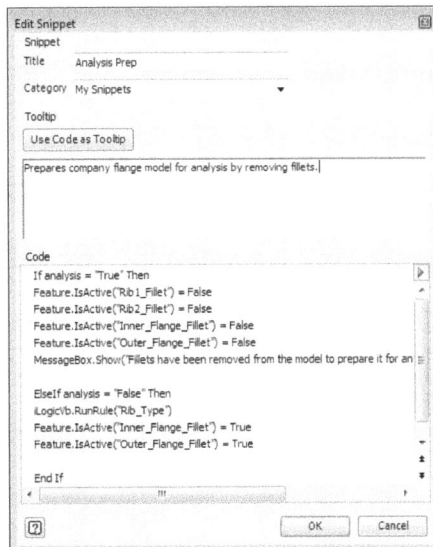

Figure 3–13

7. Click [icon] or [icon] to save the custom snippet to the default .XML file or to a new .XML file, respectively.

3.3 Rule Creation Workflow

Previously, an overall workflow was provided to explain the process of working with iLogic, starting with preparing a model to successfully completing a rule. Here the focus is on the second step of the workflow process (i.e., Rule Creation).

A rule is a Visual Basic (VB.NET) program that is embedded in a model to control its parameters, features, or components. This functionality is in addition to the user-defined parameters used for the models. iLogic rules are based on conditional statements and functions that capture the design intent, enabling you to reuse designs to meet various design scenarios. Rule creation is accomplished using the Edit Rule dialog box.

General Steps

Use the following general steps to create a new iLogic rule:

1. Create a rule.
2. Add a conditional statement.
3. Add a function.
4. Add additional functions or conditional statements, as required.
5. Terminate all conditional statements.
6. Save the rule.
7. Verify the rule.

Figure 3–14 shows the steps in Rule Creation.

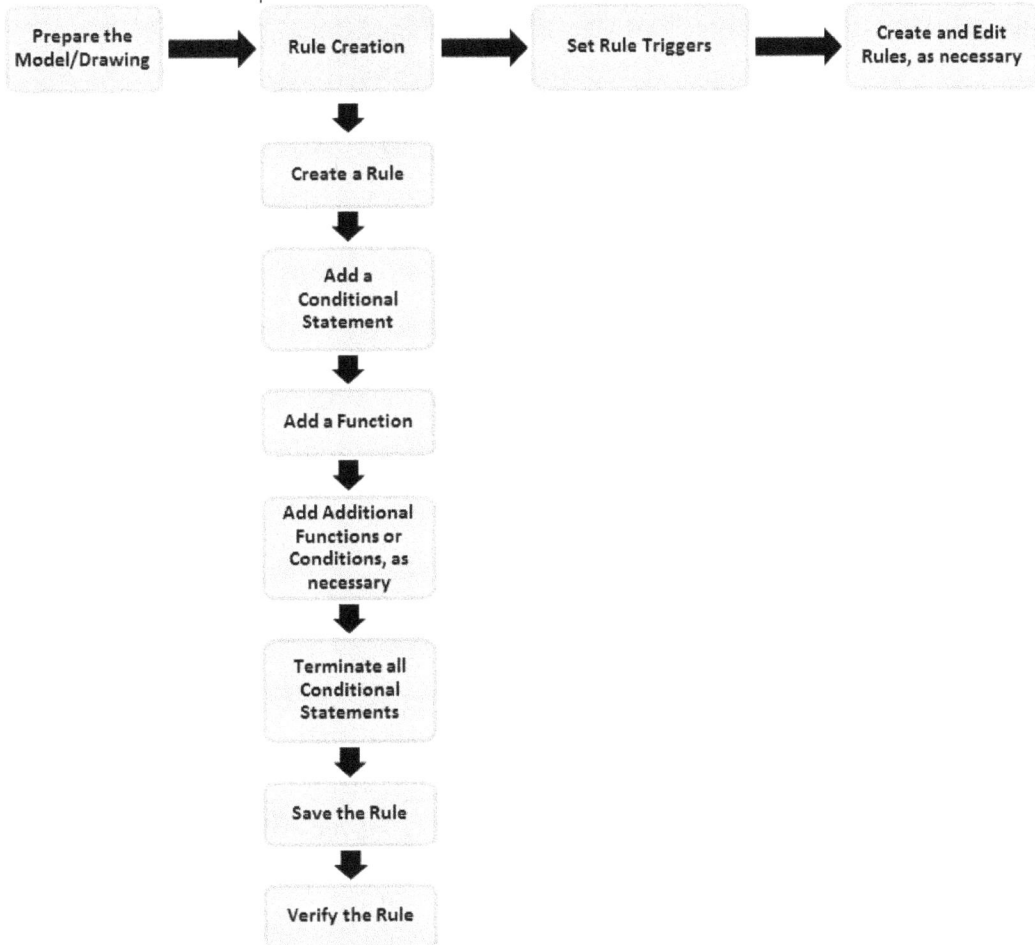

Figure 3–14

Step 1 - Create a rule.

To create a rule, in the *Manage* tab>iLogic panel, click ▤ (Add Rule), or in the iLogic browser, right-click on the model name or in the white space, and select **Add Rule**. You are prompted to enter a name for the rule and the Edit Rule dialog box opens.

- The name of the rule displays in the heading of the dialog box, as shown in Figure 3–15.

- Use descriptive and unique names for all rules to help identify their purpose in the model.

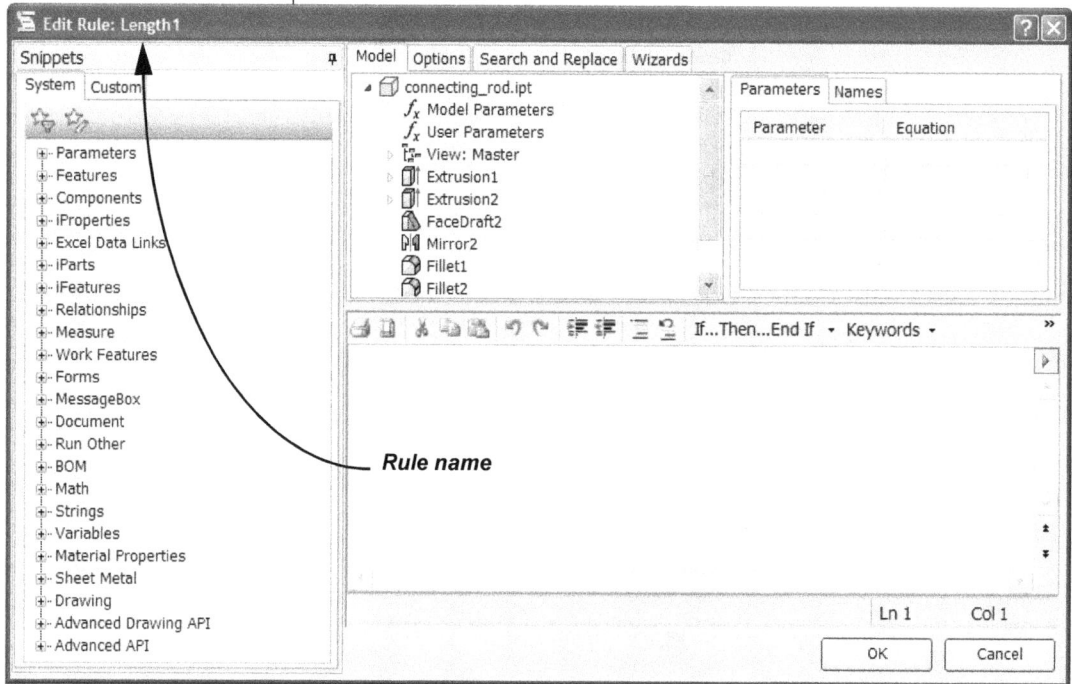

Figure 3–15

Step 2 - Add a conditional statement.

Select a conditional statement to form the basis of the rule. Conditional statements enable you to select alternate actions depending on the outcome of the condition.

- The most common conditional statement is the "If-Then-End If" statement. However, variations of this (including If, If-Then, or If-Then-Else If) can also be used.

To add a conditional statement to a rule you can do either of the following:

- Select a statement in the conditional statements drop-down list in *Rule Editor* area, as shown in Figure 3–16.

Figure 3–16

When typing a conditional statement into the Rule Editor, verify that you have the correct syntax. Conditional Statement selection is recommended.

- Type the conditional statement into the Rule Editor.

Figure 3–17 displays an "If-Then-End If" statement that has been selected in the drop-down list as the starting point for the rule. The conditional statement is displayed in red and the parameter and value placeholder (**My_Expression**) is displayed in brown.

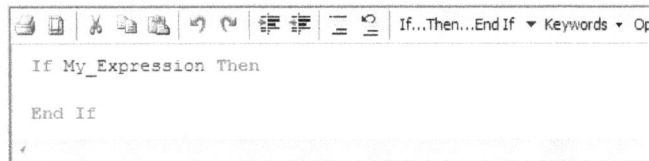

Figure 3–17

Following any "If" statement, a parameter, operand, and value are required to evaluate the initial condition of the rule. This replaces the **My_Expression** placeholder, as shown in Figure 3–18.

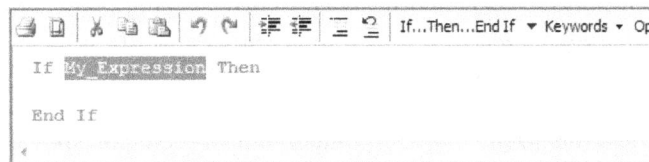

Figure 3–18

Similar to entering a conditional statement, the parameter statement can be typed or selected. To select and insert a parameter statement, select the *Model* tab in the top right corner in the Edit Rule dialog box. This area in the dialog box provides direct access to all of the model information. The *Model* tab lists all of the features, model, and user parameters in the model. As you select items in the Model list, the associated parameter or feature name become available in the *Parameters* and *Names* tabs, for selection and inclusion in the rule.

To add this information to a rule, place the cursor in the correct location in the rule and use one of the following techniques:

- Double-click on the parameter name to copy it to the rule.

- Right-click on the parameter name and select **Capture Current State** to insert both the parameter name and its value.

In Figure 3–19, the Extrusion1 feature is selected in the browser and all of its parameters are displayed. Double-clicking on the **length** parameter replaces **My_Expression** that has been selected in the Rule Editor.

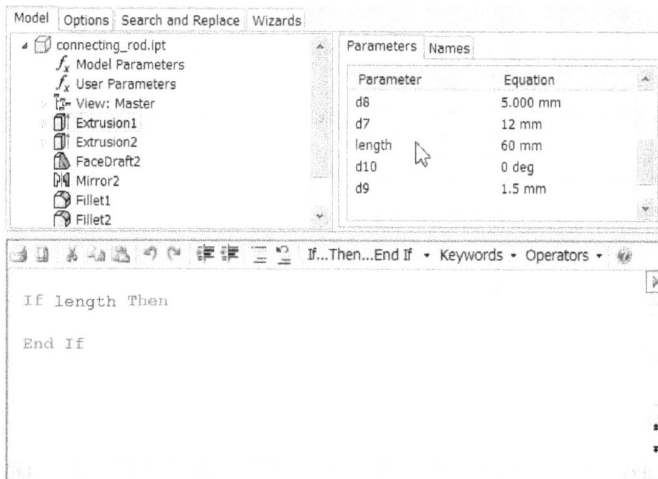

Figure 3–19

The Operators drop-down list in the row of editor tools, provides a list of common operators that can be used in a rule (e.g., <, >, <=, >=, And, Or, etc.).

You can then select an Operator in the drop-down list in the Rule Editor, or type an operator and enter a value. Once inserted, the parameters display in blue and the values display in black. You can continue to edit the conditional statement to capture the required intent. Figure 3–20 shows the "If-Then-End If" statement with the parameter, operand, and value requirements assigned.

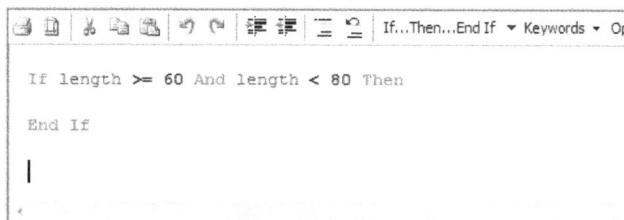

Figure 3–20

Parameter and feature names are case-sensitive. If you are typing the information in the rule, verify that the case is correct. Otherwise, the rule does not work correctly.

Step 3 - Add a function.

Immediately after the "If-Then" statement and on the line before the "End If" statement, enter the function to be performed if the specified condition is met. A list of available functions is provided in the *Snippets* area in the dialog box. To assign a snippet:

selecting a snippet ensures correct syntax and avoids unnecessary errors.

- Expand the required category and double-click on the required function to add it to the rule.

- Manually enter the function in the rule.

In Figure 3–21, the **Parameter (Dynamic)** function has been added to the rule.

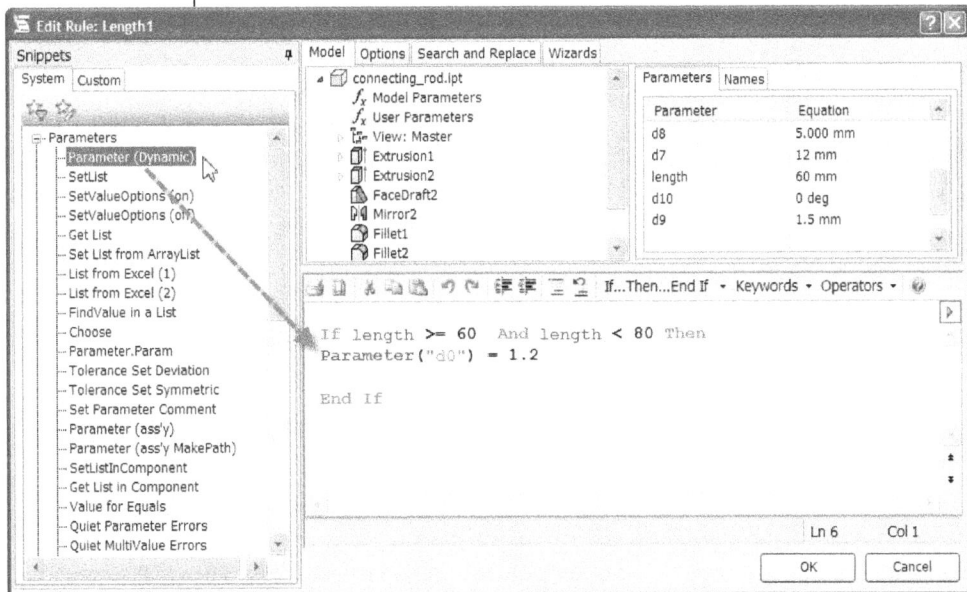

Figure 3–21

By default, the snippet displays with default parameter names and values that must be modified to capture the rule's intent.

- The parameter entry in a snippet can be typed or selected in the *Model* tab, similar to when defining the initial criteria for the condition.

For example, in Figure 3–22, if the length condition is met, the **Thickness** parameter should be equal to **3.0**.

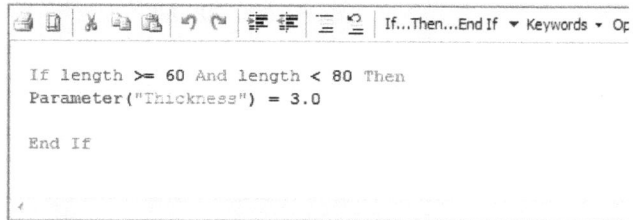

```
If length >= 60 And length < 80 Then
Parameter("Thickness") = 3.0

End If
```

Figure 3–22

Parameters vs. Variables

Parameters define model geometry or are created in the Parameters dialog box as a user-defined parameter. Both are used in iLogic to help automate a design. In addition to parameters, iLogic also uses variables. Variables are similar to a user parameter because they are not used directly in the model. However, they are used in iLogic rules to store information that is later referenced in the rule. Unlike parameters, variables are only temporary to the current iLogic rule.

Step 4 - Add additional functions or conditional statements, as required.

Continue adding functions to the rule to define the entire intent for the first conditional statement.

Not all of the rules require an "Else If" statement.

Statements can also be included to specify what happens if the initial condition is not met. Press <Enter> after the last function to add a new line. Select **Else If ... Then** in the conditional statement drop-down list. Using the same technique for defining the initial condition, define the parameter, operand, and value that is required to evaluate the second condition of the rule. For example, in Figure 3–23 an **Else If ... Then** statement was added to provide an alternate value for the **Thickness** parameter if the length was outside the initial range.

Additional conditional statements and functions can also be included after an "End If" statement, if required in the rule.

```
If length >= 60 And length < 80 Then
Parameter("Thickness") = 3.0

Else If length >= 80 And length <= 100 Then
Parameter("Thickness") = 6.0

End If
```

Figure 3–23

Step 5 - Terminate all conditional statements.

All "If" conditional statements must be terminated with an "End If" statement. If you selected a condition in the drop-down list, the "End If" statement is automatically added. If you typed the condition, enter **End If** to terminate the rule.

Step 6 - Save the rule.

Click **OK** to complete the rule and close the Edit Rule dialog box.

- If there are any errors in the rule, a dialog box opens (similar to the one shown in Figure 3–24), providing general information about the failure. Click **OK** to return to the Edit Rule dialog box to resolve the problem.

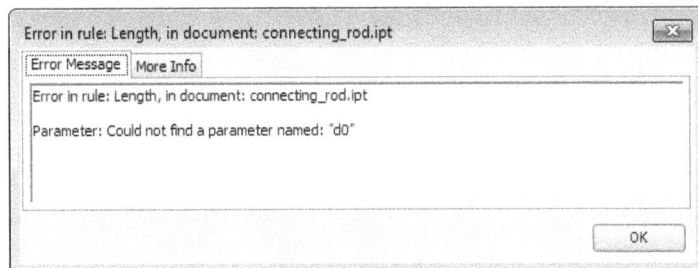

Error in rule: Length, in document: connecting_rod.ipt

Error Message | More Info

Error in rule: Length, in document: connecting_rod.ipt

Parameter: Could not find a parameter named: "d0"

OK

Figure 3–24

- If additional investigation is required outside of that provided by the Edit Rule dialog box, place the cursor in the problematic line and click ☰ (Comment) to temporarily comment out the line in the program to help troubleshoot the error.

- As an alternative to commenting out the text to prevent the error, you can click **Cancel** in the Edit Rule dialog box. The error remains in the Rule dialog box and will fail again the next time it is run, until it is resolved.

Step 7 - Verify the rule.

Modify the parameter values used to drive any of the rules to verify that the model reacts correctly when dimensional changes are made.

Rule Execution

Situations in which a rule is executed or run in a model are as follows:

- If the **Don't run automatically** option is selected in the *Options* tab for the rule, rules are run automatically when any parameter values used as variables in the rule are modified. The parameter values can be modified in the Parameters dialog box or by editing the feature to which it belongs in the model.

- A rule is executed immediately after it is created or edited, unless it is suppressed.

- Conditions in the rule that force a model change are not displayed in the model until after the rule runs. If required, you can force parameter changes to take effect immediately using specific functions in the rule (e.g., **Parameter**).

Triggers are discussed later in this student guide.

- Rules can also trigger other rules in the model.

Practice 3a

Review, Run, and Create Rules

Practice Objectives

- Identify model parameters as key parameters in the Parameters dialog box and filter them based on this setting.
- Add a new multi-value user parameter.
- Review an existing iLogic rule and identify the color coded components of the rule.
- Modify the value of a model parameter and review how its inclusion in a rule changes the model geometry.
- Add a new rule to a model that specifies the material type based on the value of a multi-value parameter.
- Load a custom snippet file into the Edit Rule dialog box.
- Add a new rule to a model using a custom snippet of code.

In this practice, the model that you open contains some renamed model parameters and a single iLogic rule. The iLogic rule controls the model geometry based on the length of a connecting rod. You will modify the **length** parameter and investigate how the rule works. Additionally, you will add a user parameter and create new rules to control the material and value of a custom parameter. The model used in this practice is shown in Figure 3–25.

Figure 3–25

Task 1 - Open a part file and edit the Parameters dialog box.

In this task, you will work in the Parameters dialog box to review the parameters that already exist in the model. While in the Parameters dialog box, you will mark specific model and user parameters as Key and filter them using this specification. Additionally, you will create a multi-value text-based user parameter that will be used in a future rule.

1. Open **connecting_rod.prt**.

2. In the *Manage* tab>Parameters panel, click f_x (Parameters).

3. In the *Key* column, select the checkbox for all of the renamed parameters (**sm_hole_inner_dia**, **lrg_hole_inner_dia**, **sm_hole_outer_dia**, **lrg_hole_outer_dia**, and **length**).

Alternatively, you can set the filter to only display the renamed parameters.

4. In the lower left corner in the Parameters dialog box, click (Filter) and select **Key**. Only the Key parameters are displayed, as shown in Figure 3–26.

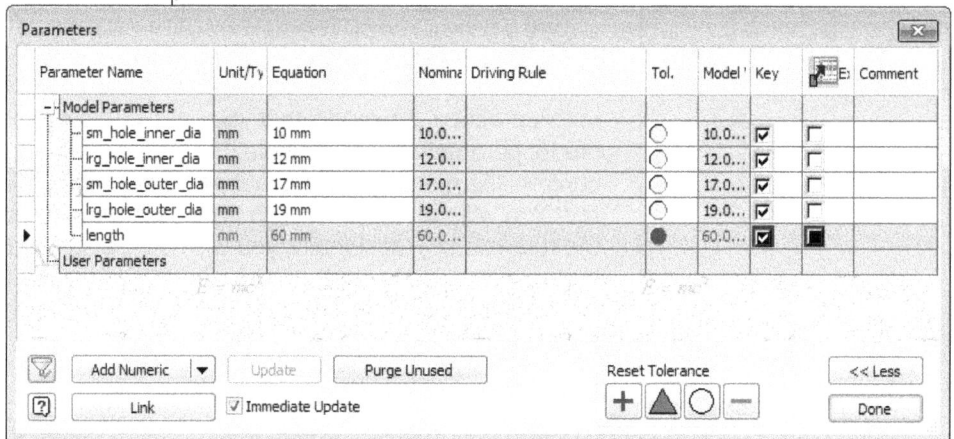

Figure 3–26

Currently there are only model parameters in the model. They were generated automatically when the geometry was created and were renamed. There are no equations in the model. In the following steps, you will create a user parameter that provides a multi-value list for a text parameter that enables for a selection of two possible styles.

5. Click **Add Text** to add a text user parameter in the **User Parameters** node in the dialog box.

6. For the name of the parameter, type **Style** and press <Enter>.

7. Right-click in any of the cells in the **Style** parameter's row and select **Make Multi-Value**. The Value List Editor displays.

8. In the *Add New Items* area, enter the values for the two Style options, as shown in Figure 3–27.

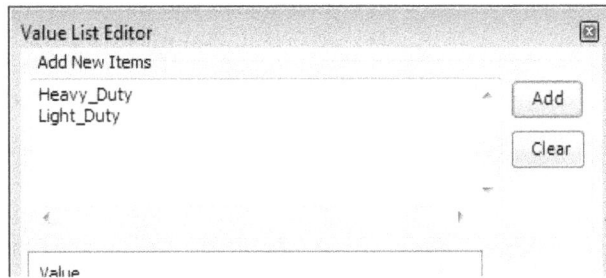

Figure 3–27

9. Click **Add**. The values are all added to the lower area.

10. Click **OK**.

11. In the *Key* column, select the checkbox for the **Style** parameter. If you do not mark the parameter as Key, the next time you open the Parameters dialog box it will not be displayed because the filter is currently set to **Key**.

12. Expand the Style drop-down list, as shown in Figure 3–28. Either Style value can be assigned.

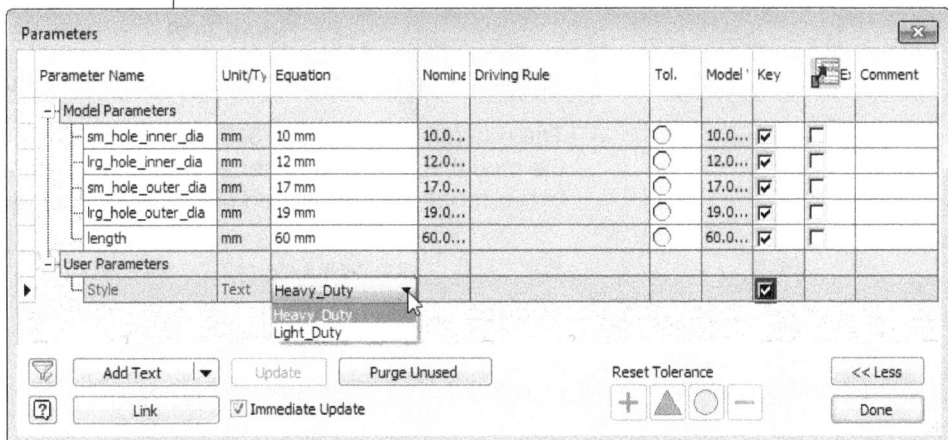

Figure 3–28

13. Close the Parameters dialog box.

Task 2 - Review an existing Rule in the model.

In this task you will review the list of iLogic rules. Using the iLogic browser, you will edit an existing rule to review the color-coded components of the rule. Once reviewed, you will modify the parameter value that is used to drive the model geometry when the rule is executed.

1. In the *Manage* tab>iLogic panel, click ᥄ (iLogic Browser) to toggle on the display of the iLogic browser, if not already on.

2. In the iLogic browser, in the *Rules* tab, one rule (i.e., Length) already exists in the **connecting_rod** model, as shown in Figure 3–29.

Figure 3–29

3. Right-click on the Length rule and select **Edit Rule**.

4. The rule (shown in Figure 3–30) was previously created in the model. It drives the inner and outer diameters of the ends of the connecting rod based on the length of the rod.

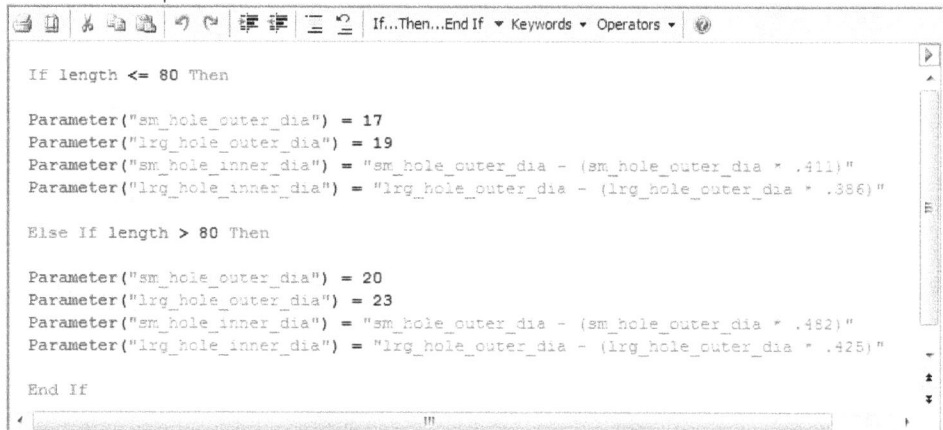

```
If length <= 80 Then

Parameter("sm_hole_outer_dia") = 17
Parameter("lrg_hole_outer_dia") = 19
Parameter("sm_hole_inner_dia") = "sm_hole_outer_dia - (sm_hole_outer_dia * .411)"
Parameter("lrg_hole_inner_dia") = "lrg_hole_outer_dia - (lrg_hole_outer_dia * .386)"

Else If length > 80 Then

Parameter("sm_hole_outer_dia") = 20
Parameter("lrg_hole_outer_dia") = 23
Parameter("sm_hole_inner_dia") = "sm_hole_outer_dia - (sm_hole_outer_dia * .482)"
Parameter("lrg_hole_inner_dia") = "lrg_hole_outer_dia - (lrg_hole_outer_dia * .425)"

End If
```

Figure 3–30

Note that the red font indicates the conditional statements in the "If..Then.. Else If.. Then" statement. The blue font indicates parameter names, the black font indicates the values, the purple font indicates functions, and the green font indicates arguments. In this rule, there are two "If" conditions. If the length is <= 80 mm, specific parameter values are used and one set of equations is used to specify the inner diameter on the inner holes. If the length is > 80 mm, other parameter values and equations are used. In general, this rule increases the amount of material in the end if the length is above 80 mm.

5. Click **OK** to close the Edit Rule dialog box.

6. In the *Manage* tab>Parameters panel, click f_x (Parameters). Note that the initial value for the **length** parameter is 60 mm. The function that is currently driving the geometry is the <= 80 mm condition.

7. In the *Equation* cell for the **length** parameter, type **90** as the new value. Note that the values for **sm_hole_inner_dia**, **lrg_hole_inner_dia**, **sm_hole_outer_dia**, and **lrg_hole_outer_dia** update in the dialog box.

8. Close the Parameters dialog box.

9. In the Quick Access Toolbar, click 🖾 if the model has not already updated. The dimensions for the model are shown in Figure 3–31.

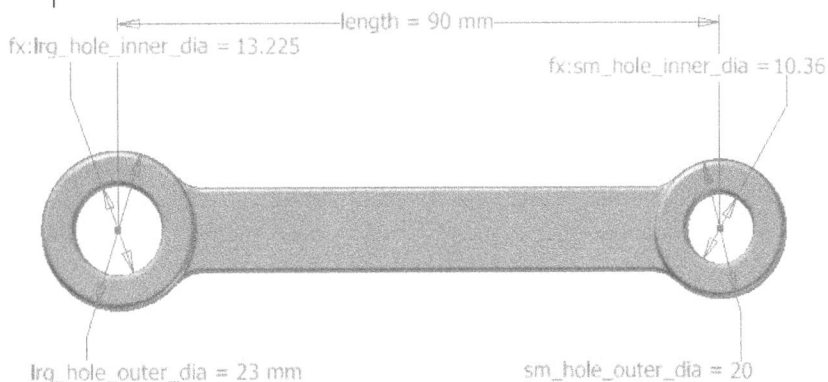

length = 90 mm

fx:lrg_hole_inner_dia = 13.225

fx:sm_hole_inner_dia = 10.36

lrg_hole_outer_dia = 23 mm

sm_hole_outer_dia = 20

Figure 3–31

10. In the *Manage* tab>Parameters panel, click f_x (Parameters).

11. Change the **length** parameter value to **55 mm**. Close the Parameters dialog box and update the model, if required. The dimensions for the ends of the connecting rod change because the functions in the <= 80 mm condition are now used. The dimensions for the model are shown in Figure 3–32.

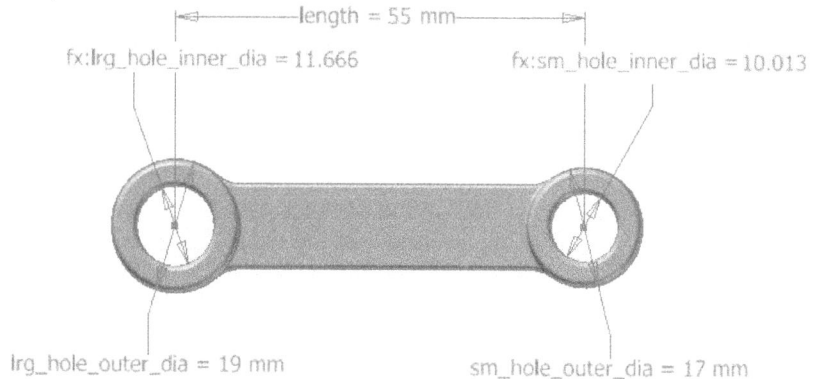

Figure 3–32

Task 3 - Add a rule to the model that controls the material.

In this task, you will add a rule to the model that will change the material used for the model depending on whether the **Style** parameter is set to **Heavy_Duty** or **Light_Duty**.

1. Before creating the Rule, review the current material that is assigned to the model. The current material is listed in the *Material* field in the Quick Access Toolbar, as shown in Figure 3–33. It is currently set as the **Generic** material with the **Default** color.

Figure 3–33

2. In the *Manage* tab>iLogic panel, click 📜 (Add Rule).

3. Enter **DutyType_Material** as the name of the rule. The Edit Rule dialog box opens.

4. In the bottom right area in the dialog box, click **If...Then...End If** to form the basis of the conditional rule.

5. At the top of the Edit Rule dialog box, in the *Model* tab, select the **User Parameters** node. All of the user parameters that have been added to the model are listed to the right, in the *Parameters* tab.

6. In the editor, highlight **My_Expression**. Double-click on the **Style** parameter that is listed in the *Parameters* tab to retrieve it into the rule. Alternatively, you can type the parameter name.

7. Type **= "Heavy_Duty"** as the remainder of the conditional statement, as shown in Figure 3–34. Note the color-coding of the statement. Conditional statements are red, parameters are blue, and the argument is green.

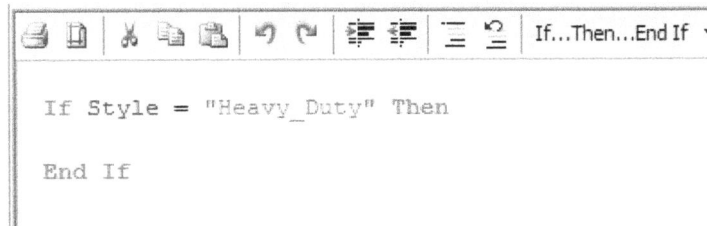

Figure 3–34

The next line in the rule must state what happens if the condition is met. In this case, if **Style = Heavy_Duty**, the material is to be set as **High Strength Low Alloy Steel**.

8. Position the cursor between the "If" and "End If" lines.

9. In the Edit Rule dialog box, in the *Snippets* area, expand iProperties and double-click on **Material** to add it to the rule.

10. In the iProperties.Material function, type **= "High Strength Low Alloy Steel"**, as shown in Figure 3–35.

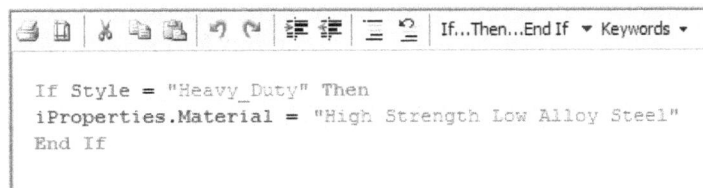

Figure 3–35

11. Place the cursor before the "End If" statement and press <Enter>.

12. Select the "Else" statement in the conditional statement drop-down list or type it. The "Else" statement displays in red, as shown in Figure 3–36.

13. Add the function as shown in Figure 3–36. You can copy and paste the function and edit the material value or add it again using the *Snippet* area in the dialog box.

```
If Style = "Heavy_Duty" Then
iProperties.Material = "High Strength Low Alloy Steel"

Else
iProperties.Material = "Carbon Steel"

End If
```

Figure 3–36

14. Click **OK** to close the Edit Rule dialog box. The Error in rule dialog box opens (as shown in Figure 3–37), indicating that the value for the *iProperties.Material* is a bad material name.

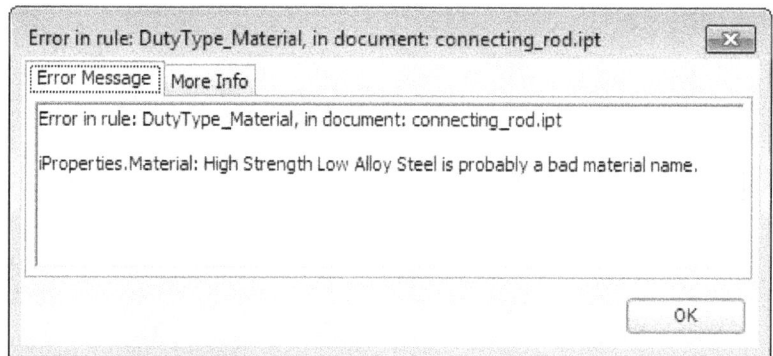

Error in rule: DutyType_Material, in document: connecting_rod.ipt

Error Message | More Info

Error in rule: DutyType_Material, in document: connecting_rod.ipt

iProperties.Material: High Strength Low Alloy Steel is probably a bad material name.

OK

Figure 3–37

15. Click **OK** to close the Error in rule dialog box. The Edit Rule dialog Box for the **DutyType_Material** rule displays.

16. Select all of the code and click ☰, as shown in Figure 3–38. This marks the selected code as commented and prevents it from being processed.

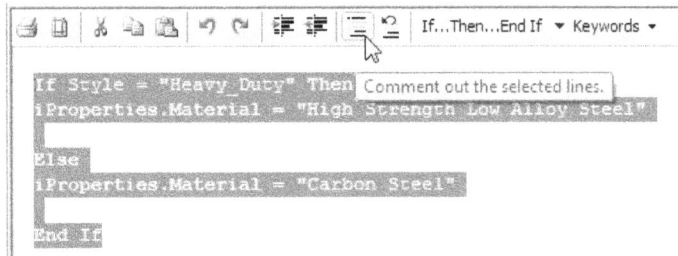

Figure 3–38

17. Click **OK** to close the Edit Rule dialog box. As the code is commented, the rule is successfully saved. You can review the correct syntax for the material types that are being used in the rule.

    As an alternative to commenting out the text to prevent the error, you can click **Cancel** in the Edit Rule dialog box.

18. In the Quick Access Toolbar, expand the Material drop-down list and scroll to the steel material types, as shown in Figure 3–39.

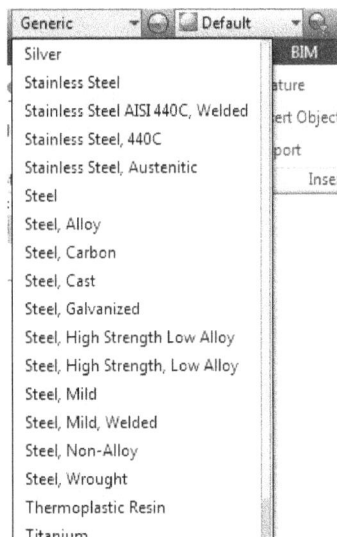

Figure 3–39

19. Note that the syntax for the material is different than what you entered in the rule. Right-click on the **DutyType_Material** rule and select **Edit Rule**.

20. Select all of the code and click ⟨⟩ to uncomment it. Edit the two function statements as shown in Figure 3–40, so that they reference the correct material names.

```
If Style = "Heavy_Duty" Then
iProperties.Material = "Steel, High Strength Low Alloy"

Else
iProperties.Material = "Steel, Carbon"

End If
```

Figure 3–40

21. Click **OK** to close the Edit Rule dialog box. Note that the Material for the model updates, as shown in Figure 3–41. (Hint: If the material value does not update correctly in the Quick Access Toolbar, right-click on connecting_rod in the Model browser and select **iProperties**. Select the *Physical* tab. The material should be assigned as **Steel,** High Strength Low Alloy. Close the iProperties dialog box and the material should update in the Quick Access Toolbar. Alternatively, you can open and close the iProperties and not use the *Physical* tab.

Figure 3–41

Task 4 - Verify the DutyType_Material rule.

In this task, verify the **DutyType_Material** rule so that the assigned material changes, based on the **Style** parameter value.

If the Quick Access Toolbar does not update, open the model's iProperties and close the dialog box to update them.

1. In the Quick Access Toolbar, click f_x (Parameters).

2. Select the *Equation* cell for the **Style** parameter and select **Light_Duty**.

3. Close the Parameters dialog box. The material updates as **Steel, Carbon**, as shown in Figure 3–42.

Figure 3–42

Task 5 - Add a custom saved snippet in a new rule.

In this task, you will add a new rule to the model that verifies that a value is assigned for the **Vendor_Name** custom parameter. This custom parameter was included in the company template file and used when the model was created. The new rule is required to verify that a value has been assigned for the **Vendor_Name** parameter. If not, you are prompted to enter a value. The rule will be created by adding a custom snippet. To access the snippet, you will load a custom .XML file from the practice files folder.

1. Right-click on the model name in the Model browser and select **iProperties**.

2. Select the *Custom* tab. Note that the **Vendor_Name** parameter exists but has not been assigned a value. Close the iProperties dialog box.

3. In the *Manage* tab>iLogic panel, click 📜 (Add Rule).

4. Enter **Vendor_Name** as the name of the rule. The Edit Rule dialog box opens.

5. Select the *Custom* tab in the *Snippets* area in the Edit Rule dialog box. Review the list of snippets. Note that the last snippet in the list is called **Print Document**.

A custom snippet file can be used when the code is reused in different models. The custom file can be loaded from a shared location and all team members can use the same file.

6. Click 📂 (Open custom snippets file). Navigate to and open *C:\Autodesk Inventor 2017 iLogic Practice Files\ iLogicClassSnippets.xml.* A new snippet is added to the end of the list that was already created.

7. Hover the cursor over the snippet to display the tooltip. It indicates that it can be used to verify whether a property is assigned to the custom Vendor Name Property.

8. Double-click on the **Vendor Name Property** snippet to add it to the Rule Editor. The rule displays as shown in Figure 3–43.

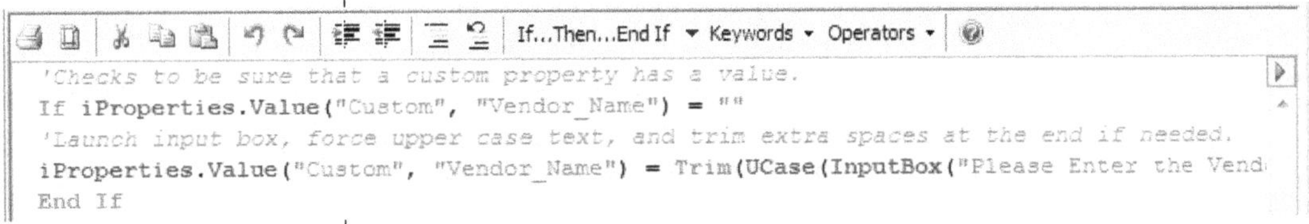

```
'Checks to be sure that a custom property has a value.
If iProperties.Value("Custom", "Vendor_Name") = ""
'Launch input box, force upper case text, and trim extra spaces at the end if needed.
iProperties.Value("Custom", "Vendor_Name") = Trim(UCase(InputBox("Please Enter the Vend
End If
```

Figure 3–43

Note that the red font indicates the conditional text of the "If..End If" statement. The blue font indicates the parameter names, the purple font indicates the functions, and the gray font indicates commented out code. In this situation, the commented code is being used to describe the purpose of the preceding lines. This is recommended when you are creating a rule to ensure that the end user understands the purpose of the code.

In this rule, there is a single condition that means it will always be executed. It checks to ensure that the **Vendor_Name** custom parameter has been assigned a value. If not, it prompts you for a value. Additionally, its sets all of the entry to be uppercase.

9. Click **OK** to close the Edit Rule dialog box.

10. The rule immediately runs and checks whether the **Vendor_Name** has a value. As it does not have a value, it prompts for the entry of a value. For the *Vendor_Name*, enter **ascent**. Click **OK**.

11. Right-click on the model name in the Model browser and select **iProperties**.

12. Select the *Custom* tab. Note that the **Vendor_Name** parameter has been assigned a value and that it in uppercase, as shown in Figure 3–44. It was originally entered as lowercase, but the rule changes it to uppercase. Close the iProperties dialog box.

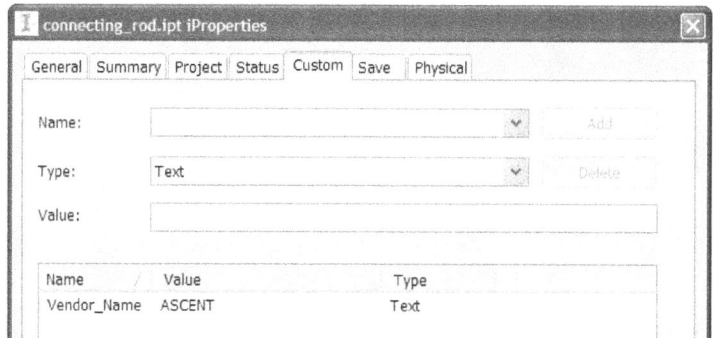

Figure 3–44

13. Right-click on the Vendor_Name rule in the iLogic browser and select **Run Rule**. Note that the rule does not launch a second time because the **Vendor_Name** parameter has now been assigned and the condition of the **Vendor_Name** parameter being empty is not being met.

14. Save and close the model.

Chapter Review Questions

1. Which of the following navigation paths provides access to the iLogic panel?

   a. *3D Model* tab>iLogic panel

   b. *Tools* tab>iLogic panel

   c. *Manage* tab>iLogic panel

   d. *Environments* tab>iLogic panel

2. Based on the color-coding used in a rule, match the color in the left column with the element type in the right column.

1. Red	a. Argument
2. Purple	b. Value
3. Black	c. Parameter
4. Green	d. Function
5. Blue	e. Condition Statement

3. The coding for an iLogic rule must be typed using a keyboard entry.

   a. True

   b. False

4. Which of the following statements is true regarding the *Snippets* area in the Edit Rule dialog box?

   a. The conditional statements that can be used in an iLogic model are listed in the *Snippets* area in the Edit Rule dialog box.

   b. The functions that can be used in an iLogic model are listed in the *Snippets* area in the Edit Rule dialog box.

   c. The operators that can be used in an iLogic model are listed in the *Snippets* area in the Edit Rule dialog box.

   d. The parameters that can be used in an iLogic model are listed in the *Snippets* area in the Edit Rule dialog box.

5. Which of the following interface components is similar to the Model browser and lists all of the views or features, and provides access to any model and user parameters that exist in the model?

    a. iLogic panel.

    b. iLogic browser.

    c. *Snippets* area in the Edit Rule dialog box.

    d. *Model* tab in the Edit Rule dialog box.

    e. Rule Editor in the Edit Rule dialog box.

6. Which of the following statements are true regarding the iLogic browser? (Select all that apply.)

    a. You can edit an existing rule using the iLogic browser.

    b. You can create a new rule using the iLogic browser.

    c. The iLogic browser display must be enabled when any model is opened.

    d. Reorder rules by dragging and dropping them in the list.

    e. Run an existing iLogic rule.

7. Match the function type in the left column with the descriptions in the right column.

1. Write	a. Functions that assign a value to a document.
2. Read	b. Functions that neither assign nor return a value.
3. Sub	c. Functions that return a value or retrieve information from a document.

8. Which of the following commands enable you to reuse existing code in an iLogic rule to create a custom snippet?

    a. ☰ (Comment Selection)

    b. ⊞ (Increase Indent)

    c. **Capture Snippet**

    d. **Save Snippet**

9. Which of the following commands enable you to create a new iLogic rule in a model?

    a.

    b.

    c.

    d.

10. Which of the following workflows best describes the process of creating a new iLogic rule?

    a. Create a Rule>Add Functions>Add Conditional Statements>Terminate Conditional Statements>Save the Rule>Verify the Rule.

    b. Create a Rule>Add Conditional Statements>Terminate Conditional Statements>Add Functions>Save the Rule> Verify the Rule.

    c. Create a Rule>Add Conditional Statements>Add Functions>Terminate Conditional Statements>Save the Rule>Verify the Rule.

    d. Create a Rule>Add Conditional Statements>Add Functions>Terminate Conditional Statements>Verify the Rule>Save the Rule.

Command Summary

Button	Command	Location
	Add Rule	• **Ribbon:** *Manage* tab>iLogic panel
NA	**Capture Snippet**	• **Shortcut:** (*right-click with code selected in the Rule Editor*)
	Comment/ Uncomment lines	• **Toolbar:** (*in the Rule Editor in the Edit Rule dialog box*)
	Copy	• **Toolbar:** (*in the Rule Editor in the Edit Rule dialog box*)
	Cut	• **Toolbar:** (*in the Rule Editor in the Edit Rule dialog box*)
	Edit favorite snippets	• **System Toolbar:** (*in the Snippets area in the Edit Rule dialog box*)
	Help	• **Toolbar:** (*in the Rule Editor in the Edit Rule dialog box*)
If...Then...End If		• **Toolbar:** (*in the Rule Editor in the Edit Rule dialog box*)
	iLogic Browser	• **Ribbon:** *Manage* tab>iLogic panel
	Increase/ Decrease Indent	• **Toolbar:** (*in the Rule Editor in the Edit Rule dialog box*)
Keywords		• **Toolbar:** (*in the Rule Editor in the Edit Rule dialog box*)
	Merge custom snippet files	• **Custom Toolbar:** (*in the Snippets area in the Edit Rule dialog box*)
	Open custom snippet file	• **Custom Toolbar:** (*in the Snippets area in the Edit Rule dialog box*)
Operators		• **Toolbar:** (*in the Rule Editor in the Edit Rule dialog box*)
	Page Setup	• **Toolbar:** (*in the Rule Editor in the Edit Rule dialog box*)
	Paste	• **Toolbar:** (*in the Rule Editor in the Edit Rule dialog box*)
	Print	• **Toolbar:** (*in the Rule Editor in the Edit Rule dialog box*)
	Save custom snippet file	• **Custom Toolbar:** (*in the Snippets area in the Edit Rule dialog box*)
	Toggle snippet filtering	• **System Toolbar:** (*in the Snippets area in the Edit Rule dialog box*)

Rule Creation

Rules consist of conditional statements and functions. The conditional statement defines which instructions are to be executed based on the outcome of a condition. The instructions that are programmed in the rule are called functions. The instructions in a function is what interacts with the model geometry. It is important to learn about the different types of conditional statements and operators that can be used in a rule along with the many types of Snippets that are available in the Edit Rule dialog box to create rules that correctly drive model geometry.

Learning Objectives in this Chapter

- Identify the required "If...Then...Else/Else If...End If" conditional statement format to form the basis of a required iLogic rule.
- Incorporate the use of Case statements as an alternative to using extended block If statements.
- Understand the mathematical operators, functions, and unit specifications that can be used in an iLogic rule.
- Add a single value Parameter function to an iLogic rule and edit its variables and syntax to correctly read or get information from the model.
- Add a multi-value Parameter function to an iLogic rule and edit its variables and syntax to correctly populate the multi-value list or read the current value from the model.
- Set the function's variable that defines which sub-component of an assembly is to be affected by the function.
- Incorporate a Feature function in an iLogic rule that controls suppression or verifies a feature's suppression state.
- Assign colors to part and assembly features using an iLogic rule.
- Assign or read thread Type, Designation, or Class Specifications for a Thread feature using an iLogic rule.
- Incorporate iProperty functions in an iLogic rule to control the iProperty values for a model.

iLogic Workflow

Figure 4–1 shows the overall suggested workflow for the iLogic tools in the Autodesk® Inventor® software. The horizontal line at the top represents the high-level workflow and each of their sub-steps are detailed vertically below them. The highlighted column represents the content discussed in the current chapter.

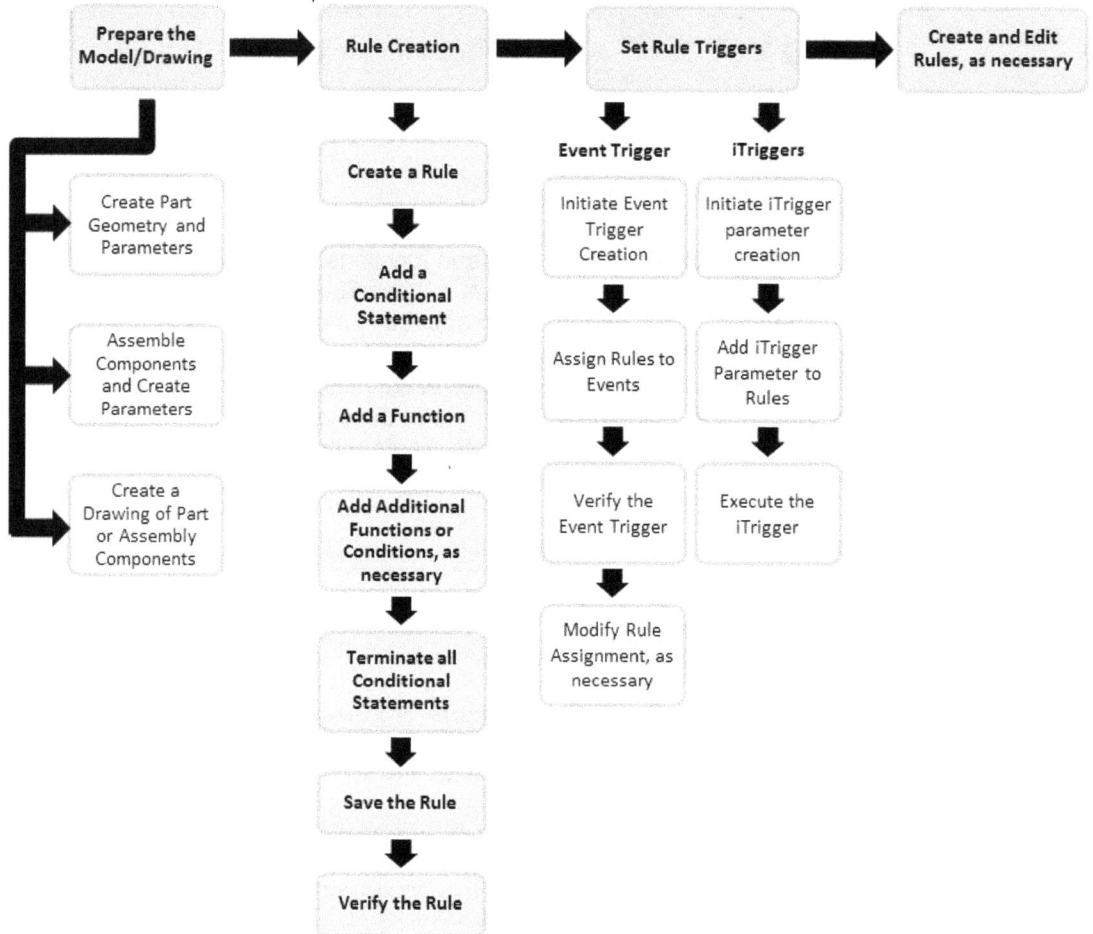

Prepare the Model/Drawing	Rule Creation	Set Rule Triggers	Create and Edit Rules, as necessary

Create Part Geometry and Parameters

Assemble Components and Create Parameters

Create a Drawing of Part or Assembly Components

Create a Rule

Add a Conditional Statement

Add a Function

Add Additional Functions or Conditions, as necessary

Terminate all Conditional Statements

Save the Rule

Verify the Rule

Event Trigger   iTriggers

Initiate Event Trigger Creation   Initiate iTrigger parameter creation

Assign Rules to Events   Add iTrigger Parameter to Rules

Verify the Event Trigger   Execute the iTrigger

Modify Rule Assignment, as necessary

Figure 4–1

4.1 Conditional Statements

When creating a rule a conditional statement is generally used to form the basis of the rule. Conditional statements enable you to select alternate actions depending on the outcome of a condition. A common conditional statement is the "If...Then...End If" statement. However, variations of this can also be used. Three standard conditional statements are included in the conditional statement drop-down list in the Rule Editor toolbar, as shown in Figure 4–2.

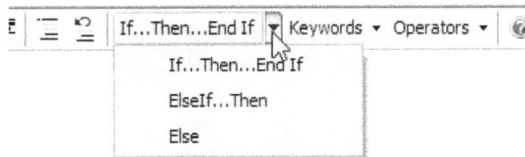

Figure 4–2

- You can select a statement or manually type it into the Rule Editor.

- In general, the If...Then...EndIf statement should be the first conditional statement that is used in a rule. It sets the If condition and the action that is executed. The other statements can be combined with this to define further conditions and actions.

- Conditional statement elements are displayed as red in a rule.

Conditional statement selection is recommended to ensure you are using the correct syntax.

If...Then...EndIf

In an If...Then...EndIf statement, there is a single outcome based only on whether the condition is true or not. If the condition is false, the conditional statement is skipped entirely and no action is taken. The following example shows the use of an If...Then...EndIf statement.

```
If overall_length = "large" Then
d0 = 36 in
End If
```

If...Then...Else...EndIf

In an If...Then...Else...EndIf statement, there are two possible outcomes. If the first condition is true, its actions are executed. If the condition is false, another set of actions are executed. This type of statement cannot be explicitly selected in its entirety with a single selection in the conditional statement drop-down list; however, you can create it using the combination of statements. Begin by defining the If...Then portion of the statement, then place the cursor before the EndIf statement, and then select the Else statement in the conditional statement drop-down list. Alternatively, manually type Else statement and then add its action. The following example shows a use of an If...Then...Else statement.

```
If overall_length = "large" Then
d0 = 36 in
Else
d0 = 12 in
End If
```

If...Then...ElseIf...Then...EndIf

This extended block If statement is a combination of the If...Then...EndIF and ElseIf...Then statements that are listed in the conditional statement drop-down list. In combination, it enables for multiple conditional statements with multiple outcomes. If the first condition is true, its actions are executed. If the second condition is true, another set of actions are executed, etc. To complete the statement, it can have an Else statement to take into account a false result or the statement can simply end and the rule does not generate an action if no conditions are met. It is shown in the following example.

```
If overall_length = "large" Then
d0 = 36 in
ElseIf overall_length = "medium" Then
d0 = 24 in
ElseIf overall_length = "small" Then
d0 = 12in
End If
```

If

If statements can also be written as single lines. In this situation, no EndIf statement is required. The following example shows a use of a single line If statement.

For example:

```
If overall_length = "large" Then d0 = 36 in
```

Case Structure

As an alternative to using an extended block If statement, you can use Case structure to drive a rule that has more than two alternatives. The programming elements for a Case Structure statement are located in the **Keywords** down-down list in the Rule Editor toolbar, as shown in Figure 4–3.

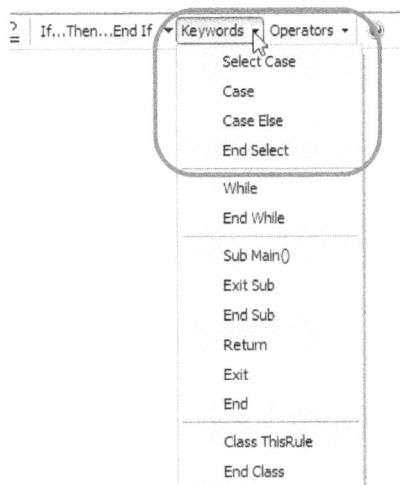

Figure 4–3

In general, a Case structure rule begins with the Select Case statement followed by a parameter (variable) that is to be checked. Case statements are then added to describe the action if the parameter or variable is equal to the defined value. There are a number of different ways to write a Case statement. These include:

- Writing multiple Case statements to include a single required value for the parameter. The value is included in quotes. The following example shows this type of Select Case statement.

    ```
    Select Case overall_length
    Case "large"
    d0 = 36 in
    Case "medium"
    d0 = 24 in
    Case "small"
    d0 = 12 in
    End Select
    ```

- Writing the Case statement to include a final statement that indicates what happens if an unexpected value is read. The following example shows the use of this type of Select Case statement, where if the value for the **overall_length** parameter is not large, medium, or small, the value of **d0** is forced to 12 in (small).

*A Case Else statement can also include the delivery of a message box that prompts to re-enter the **overall_ length** using an appropriate value.*

```
Select Case overall_length
Case "large"
d0 = 36 in
Case "medium"
d0 = 24 in
Case "small"
d0 = 12 in
Case Else
d0 = 12 in
End Select
```

The conditional statement for a Case statement can include Operators such as AND, OR, etc.

- Writing Case statements that include multiple or a range of required values for the parameter. The following example shows these types of Select Case statement.

```
Select Case length
Case 1, 2, 3, 4
width = 4 in
Case 4 to 10
width = 10 in
Case Is > 10
width = 12 in
End Select
```

While

A While...End While statement enables you to create a repeatable block that runs as long as a condition is true. A while statement can also have an optional Else statement. This statement can be manually added to a rule or it can be selected from the **Keywords** pull-down list in the Rule Editor toolbar.

- If an endless loop is encountered in the While...End While statement, iLogic is unable to process the loop. You must manually terminate the Autodesk Inventor process in order to exit it.

- Consider adding extra lines of code to the rule to define all conditions that are to be tested instead of using a While...End While statement that might develop as an endless loop.

Boolean Variables in Rules

A boolean variable holds a True or False value. Boolean variable statements can be used in an iLogic program to control output of a rule. For example, if a feature is suppressed in a model, its boolean variable is False. If it is displayed, it is True. The use of boolean variables can be used in iLogic rules. Boolean parameters can also be created in the Autodesk Inventor software using a True/False user parameter.

The Feature.IsActive function allows for the suppression or display of a feature in a model.

- To write a conditional statement that checks whether Fillet1 is active or not, you can use either of the following statements which include the use of a function that defines the suppression state. In the second case, the true value is implied and in the fourth case, the false value is implied.

  ```
  If Feature.IsActive ("Fillet1") = true Then....
  If Feature.IsActive ("Fillet1") Then....
  If Feature.IsActive ("Fillet1") = false Then....
  If Not Feature.IsActive ("Fillet1") Then....
  ```

- A conditional statement can also be written that references the boolean value of a True/False user parameter as shown in the following example.

  ```
  If fillet = True Then
  Feature.IsActive ("Fillet1") = True
  Else
  Feature.IsActive ("Fillet1") = False
  End If
  ```

No Action Required

Conditional statements can also be written when no action is taken if a certain condition is met. In this situation, you can begin with the If...Then...EndIf statement and when writing the condition in which no action is required, you can enter a comment such as 'No Action Required'. This helps to identify that no action is taken in this situation. An Else statement can also be added to further define the conditional statement, if required.

```
If width < = 20 Then 'No Action Required'
Else
width = 20
MessageBox.Show("Invalid Size. A width greater than 20 has been
entered. The width has been set to 20.", "Size Error")
End If
```

4.2 Operators

Operators are the iLogic program elements that are used to test conditions or define values. Most standard programming operators are available for use in iLogic rules. Additionally, a predefined list of commonly used operators is available, as shown in Figure 4–4. This list is accessed from Operators drop-down list in the Rule Editor toolbar in the Edit Rule dialog box. You can select an operator from this list or you can manually type it into a rule.

Additional mathematical operators are located in the Math category in the System tab in the Snippets area (e.g., Sin, Cos, Ln, etc.).

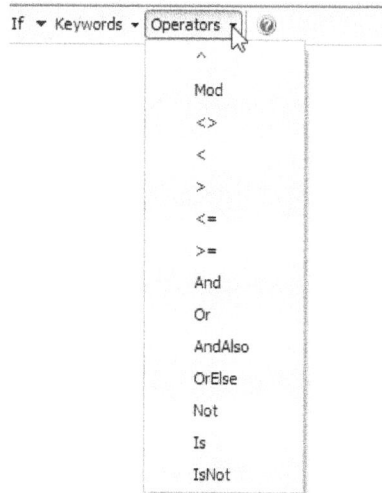

If ▼	Keywords ▼	Operators

^
Mod
<>
<
>
<=
>=
And
Or
AndAlso
OrElse
Not
Is
IsNot

Figure 4–4

Mathematical Operators and Functions

In addition to the provided list of operators, the following mathematical operators can be manually typed and used in rules.

+	Addition
-	Subtraction
/	Division
=	Equal to
*	Multiplication
^	Exponentiation
()	Expression delimiter

Mathematical functions can also be manually typed into a rule or you can select them from a predefined list. To access this list, right click in the Rule Editor and expand **Math Functions**, as shown in Figure 4–5.

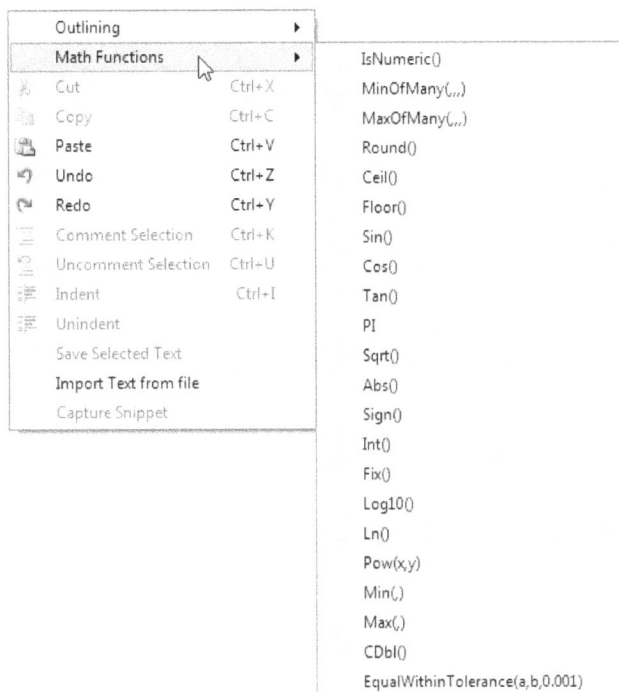

Outlining	▶	
Math Functions	▶	IsNumeric()
Cut	Ctrl+X	MinOfMany(,,,)
Copy	Ctrl+C	MaxOfMany(,,,)
Paste	Ctrl+V	Round()
Undo	Ctrl+Z	Ceil()
Redo	Ctrl+Y	Floor()
Comment Selection	Ctrl+K	Sin()
Uncomment Selection	Ctrl+U	Cos()
Indent	Ctrl+I	Tan()
Unindent		PI
Save Selected Text		Sqrt()
Import Text from file		Abs()
Capture Snippet		Sign()
		Int()
		Fix()
		Log10()
		Ln()
		Pow(x,y)
		Min(,)
		Max(,)
		CDbl()
		EqualWithinTolerance(a,b,0.001)

Figure 4–5

Units

Units are assigned to each parameter value. If you do not assign them, the software can usually resolve equations that use simple operators (+, -, *, /). To ensure that the correct dimension type is used, it is recommended that you enter the value with its unit.

Some of the symbols used for units of distance are as follows:

When using unit symbols in a rule, it is recommended that you refer to the Unit Specification dialog box to ensure the correct syntax. To access this dialog box, select any one of the Unit/Type cells in the Parameters dialog box.

Unitless	ul
Inches	in, inch, or "
Feet	ft, foot
Meter	m, meter
Centimeter	cm
Millimeter	mm

ul indicates that the value is unitless. In situations where a unit is omitted, the system assumes the measurement type. For example, if you edit an inch dimension by entering **3/2**, the software interprets the units required to evaluate the expression as 3 in/2 ul.

A value's unit specification is applied before mathematical operations are performed and have precedence over all other operators. In the text of an iLogic rule, the unit names:

- must be used directly after numbers.

- cannot be used after parameters, variables, or expressions.

- must include a space between the number and the unit.

For example, 23 in, 100 m^2, 15 m/s are valid numeric values with units. For operators such as exponents, you should assign units to each numerical value when written in a rule. For example, enter **3in^2ul** to specify a distance of the square of three inches.

To incorporate a complicated compound unit specification into a rule, it is recommended that you create an equation for it in the Parameters dialog box and assign the units using the Unit Type dialog box. To copy it into the Rule Editor, use the **Capture Current State** command on that parameter when creating the rule.

4.3 Parameter Functions

The *Parameters* category in the *System* tab, in the *Snippets* area, provides access to a list of functions that are associated with parameters in a model. The expanded list of Parameter functions is shown in Figure 4–6.

Figure 4–6

To obtain a general description and the syntax for a function in the list, hover the cursor over the function to display a tooltip. For example, the tooltip (shown in Figure 4–7) indicates that the Parameter (Dynamic) function can be used to set or get the current value for a parameter while a rule is running. The rule's syntax is listed below the help line in the tooltip.

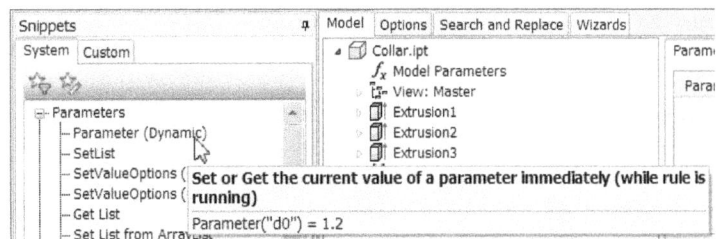

Figure 4–7

Functions that enable you to interface with API programs are not covered in this student guide.

The majority of functions in the *Parameters* category can be sub-divided into functions that deal with single or multiple value parameters. Additionally, there are some miscellaneous functions and advanced functions that deal with API programming. Only the commonly used functions are discussed. For complete information on the *Parameters* category, refer to the Autodesk Help documentation.

Parameter Functions

Parameter functions are used to assign a single value to a parameter or to read a single value from a parameter.

- They can be used in the same model as the rule or in a component in an assembly when the rule exists in the top-level assembly to assign a value. In the first example shown below, the parameter resides in the same model as the rule using the **Parameter (Dynamic)** function. In the second example, the parameter exists in a component of the assembly and the part name is required. This was written using the **Parameter (ass'y)** function. These statements can be used in a conditional statement to test a condition or in an action statement to assign a value.

        Parameter ("d0") = 22.7
        Parameter ("PartA:1","d0") = 22.7

*If a parameter cannot be found, an error message is returned. To prevent this message from displaying, use the **Quiet Parameter Errors** function above the parameter function line.*

__Parameter.Quiet = True__

- You can also use the **Parameter (Dynamic)** and **Parameter (ass'y)** functions to read a value from the model and assign it to another parameter. In the following examples, the value of **"d0"** is being assigned as the value for another parameter. In the first example shown below, the parameter resides in the same model as the rule. In the second example, the parameter exists in a component of the assembly and the part name is required.

        param_val = Parameter ("d0")
        param_val = Parameter ("PartA:1","d0")

- If a component is a member of a sub-assembly you use the following MakePath syntax to locate the model using the **Parameter (ass'y Make Path)** function.

        Parameter (Make Path("SubAssem:A:1", PartA:1"),"d0") = 22.7

MultiValue Functions

A number of functions in the *Parameters* category are multi-value type functions. These functions are used to access and change the list of values associated with multi-value parameters.

Setting and Getting Values

The **SetList** function enables you to set the list of options for a multi-value parameter. In addition to defining parameter values for the list, it can also prevent users from deleting a value in the Parameters dialog box as they are recreated when the rule is run.

Consider the use of the SetValueOptions (on) and SetValueOptions (off) functions in a program if the values in a multi-value list can be changed. They can specify which values to use in the list.

- The list can include values or equations, or a combination of two multi-value numeric parameters, and text strings for multi-value text parameter. In the examples shown below, the **SetList** function is used to assign numeric values (example 1), values based on equations (example 2), and text strings for a parameter called **Size** (example 3).

```
MultiValue.SetList("d0", 0.5, 0.75, 1.0, 1.25)
MultiValue.SetList("d0", "3/8", "d1 * 2", "d1 * 3")
MultiValue.SetList("Size", "Value1", "Value2")
```

The order values listed in the code determine the list order. The list does not adjust alpha-numerically as it does when values are assigned in the Parameters dialog box.

- Similar to assigning parameter values, you can set the value for a list in another component using the **SetListInComponent** function. In this function, you also specify the part name, as shown in the following example.

```
MultiValue.SetList("Part1:1", "d0", 0.5, 0.75, 1.0, 1.25)
```

*If a parameter cannot be found, an error message is returned. To prevent this message from displaying, use the **Quiet MultiValue Errors** function above the parameter function line.*

The **GetList** and **Get List in Component** functions enable you to get the current list value for use in a rule. In the first example below, the value returned is the current value set for the **d0** parameter. In the second example, the current value is set for the **d0** parameter in the Part1 component.

```
values = MultiValue.List("d0")
values = MultiValue.List("Part1:1", "d0")
```

MultiValue.Quiet = True

A few functions enable you to populate a multi-value parameter using defined lists in the Autodesk Inventor file or in an external file.

- The Autodesk Inventor software has a defined list of materials based on the active material library. A user parameter can be automatically populated with this list using the **Set List from ArrayList** function. In the following example, the user parameter called **Material** is populated with the complete Materials list for the currently active material library.

      MultiValue.List("Material") = iProperties. Materials

- Multi-value parameter values can also be populated from an external Microsoft Excel file using the **List from Excel (1)** and **List from Excel (2)** functions. The Excel (2) function has fewer required fields and can only be used in a rule where the Excel (1) function has been already used and the filename and sheet have been defined.

      MultiValue.List("d0") = GoExcel.CellValues ("filename.xls, "Sheet1", "A2", "A10")
      MultiValue.List("d0") = GoExcel.CellValues ("B2", "B10")

Find a value

Values can be located and reused in a rule. The values can be pulled directly from a multi-value list for a specified parameter. The syntax for the function indicates conditions on the value that is to be returned. The condition on the returned value can use the "<=", "=", or ">=" operators. In the following example, the multi-value parameter list for the **d0** parameter is searched and the closest value that is < or = is found and returned.

      foundVal = MultiValue.FindValue (MultiValue.List("d0"), "<=", 4.0)

Miscellaneous Parameter functions

*The **iLogicVb.Update WhenDone = True** function can be used in a rule to force the entire model to update once the function is encountered. The **UpdateAfterChange** and **UpdateAfterChange for MultiValue** options are for custom dialog boxes. This is not covered in this student guide.*

- The comment associated with a parameter or multi-value parameter can be a populated using the **Set Parameter Comment** function. This function can be used to explain the parameter or to describe that the parameter is driven by a rule. In the following example, the comment for the **Material** parameter is *List populated using the Material Library List.*, providing information on how the multiple values were added.

    ```
    Parameter.Param("Material").Comment = "List populated using the
    Material Library list"
    ```

- When using any of the Parameter functions that have been discussed, the Parameter must exist in the model before the function is used. The **Load XML** function enables you to read an XML file to populate the user parameter fields with new parameters and values. The **Save XML** and **Save XML (keys only)** functions enable you to export an XML file from the current model that contains the parameter information.

4.4 Feature Functions

The *Features* category in the *System* tab, in the *Snippets* area, provides access to a list of functions that deal specifically with features. These enable you to set or read feature suppression states, colors, and thread designations. The expanded list of Features functions is shown in Figure 4–8.

To obtain a general description and the syntax for a function in the list, hover the cursor over the function to display a tooltip describing its purpose and syntax.

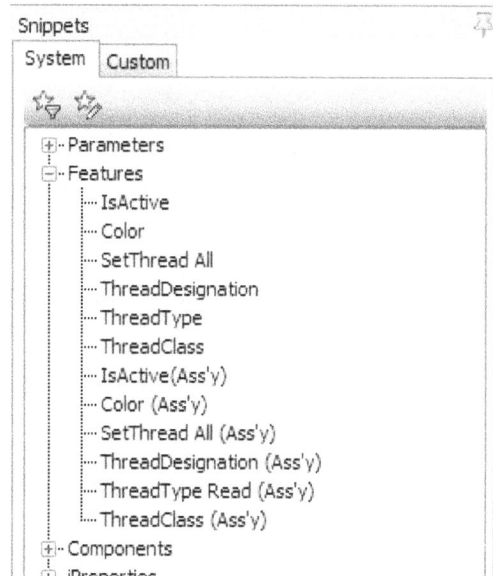

Figure 4–8

The *Features* category of functions can be sub-divided into functions that control feature suppression states, color, and threads properties. For additional information on the *Features* category of functions, refer to the Autodesk Help documentation.

Feature Suppression States

The **IsActive** function can be used to either set or read the suppression state of a feature. Because of its ability to both set and read, it can be used as a test condition in an If or ElseIf statement or it can be used as a function that is to be executed if a condition is met. You can also incorporate both situations in a single rule. The **IsActive (Ass'y)** function has the same purpose as the IsActive function; however, it is used to control features in subcomponents of an assembly using rules in a top-level assembly and therefore, provides an extra field for locating the file. Examples of code using these functions are as follows:

- In the following example, the **IsActive** function is activated if a conditional statement is met in order to control the suppression state of features in the model. Specifically, the rule displays Rib1 and suppresses Rib2 if the **number_of_ribs** parameter has a value of less than or equal to four. If the **number_of_ribs** is greater than 4, Rib1 is suppressed and Rib2 is displayed.

```
If number_of_ribs <= 4 Then
Feature.IsActive("Rib1")= True
Feature.IsActive("Rib2")= False
Else
Feature.IsActive("Rib1")= False
Feature.IsActive("Rib2")= True
End If
```

- In the following example, the **IsActive** function is used in both the conditional and action portion of a rule. The rule reads the suppression status on the Mounting_Hole feature and if it is displayed, it suppresses the Central_Hole feature. If the Mounting_Hole feature is suppressed, it displays the Central_Hole feature.

```
If Feature.IsActive("Mounting_Hole") = True Then
Feature.IsActive("Central_Hole") = False
Else
Feature.IsActive("Central_Hole") = True
End If
```

- In the following example, the same code discussed for the number of Ribs is used; however, the functions being executed are in a component (Part1) of an assembly.

```
If number_of_ribs <= 4 Then
Feature.IsActive("Part1:1, "Rib1") = True
Feature.IsActive("Part1:1, "Rib2")= False
Else
Feature.IsActive("Part1:1, "Rib1")= False
Feature.IsActive("Part1:1, "Rib2")= True
End If
```

Color

The **Color** and **Color (Ass'y)** functions can be used to either set or read the color of a feature in the same model or in a component of an assembly. Because of its ability to both set and read, it can be used as a test condition in an If or ElseIf statement or it can be used as a function that is to be executed if a condition is met. You can also incorporate both situations in a single rule.

- In the following example, the color of the Base feature changes to Red if the value for the **number_of_holes** parameter is less than 6 or is Blue if the value is greater than or equal to 6.

  ```
  If number_of_holes < 6 Then
  Feature.Color("Base") = "Red"
  Else
  Feature.Color("Base") = "Blue"
  End If
  ```

- Similar to the assembly version of other functions, the **Color (Ass'y)** function requires the additional field to direct the action to a specific component in an assembly.

  ```
  If number_of_holes < 6 Then
  Feature.Color("Part1:1", "Base") = "Red"
  Else
  Feature.Color("Part1:1","Base") = "Blue"
  End If
  ```

Thread Properties

There are a number of functions that enable you to set and read specific properties for thread or tapped hole features. These properties are those that display in the *Specification* tab in the Thread dialog box, as indicated in Figure 4–9 or in the similar dialog box when creating a tapped hole. These include Thread Type, Designation, and Class.

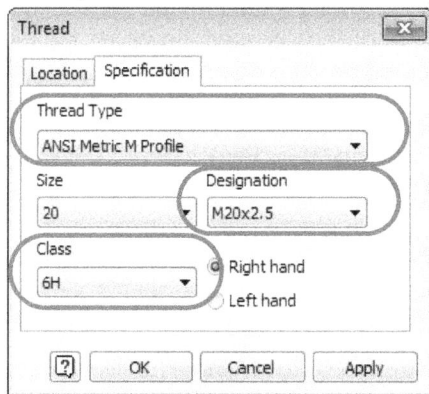

Figure 4–9

Similar to all functions that are used in assemblies, the functions that end with (Ass'y) are used similar to the part level functions; however, an additional variable that identifies the part name of a sub-component is required.

In general, the variables that are used in the thread functions are intuitive based on their name. For example, the featurename variable refers to the name of the feature to which the thread properties are being read or assigned to. Thread type, thread designation, and thread class all refer to the specification that is being read or set.

- In the following example, the **SetThread All** function is used. The thread type, designation, and class specification are all being assigned at one time to the Thread1 feature. The type is being set as ANSI Unified Screw Threads, its designation as 13/16-10 UNC, and its class as 2B.

        Feature.SetThread("Thread1", "ANSI Unified Screw Threads",
        "13/16-10 UNC", "2B")

- The **ThreadDesignation** and **ThreadClass** functions can be used individually to assign the designation or class for a thread, as shown in the following examples.

        Feature.ThreadDesignation("Thread1") = "13/16-10 UNC"
        Feature.ThreadClass("Thread1") = "2B"

*To assign a value for thread type, you must use the **SetThread All** function and assign all three properties.*

- The **ThreadType** function is a read only function and can only read the current value of the thread type. For example, you can read the current type and use it to populate the value of a parameter (i.e., **T_Type**) using the following syntax.

        T_Type = Feature.ThreadType("Thread1")

4.5 iProperty Functions

The *iProperty* category in the *System* tab, in the *Snippets* area, provides access to functions that deal specifically with the iProperties of the Autodesk Inventor part, assembly, or drawing documents. The expanded list of iProperties functions is shown in Figure 4–10.

To obtain a general description and the syntax for a function in the list, hover the cursor over the function to display a tooltip describing its purpose and its syntax.

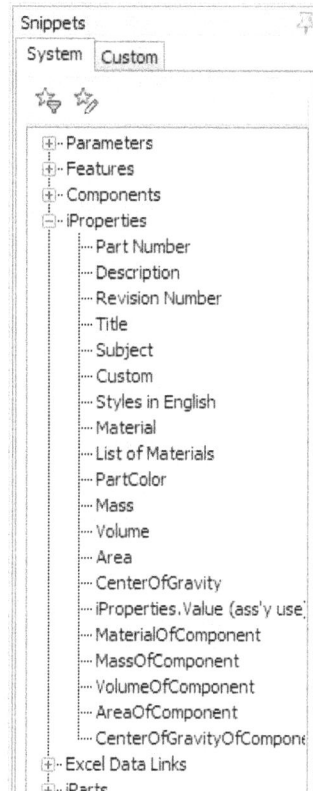

Figure 4–10

iProperty Dialog Box Properties

To access a file's iProperty data, right-click its filename in the Model Browser and select **iProperties**. The iProperties dialog box opens as shown in Figure 4–11.

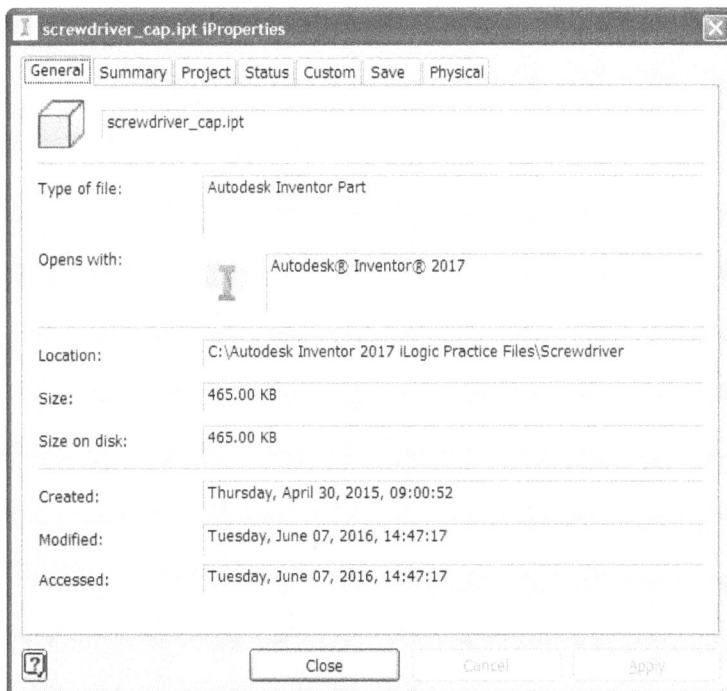

Figure 4–11

*Some iProperties support different languages. For example, materials and color. It is recommended that English names for materials and color are used. The **iProperties.StylesInEnglish = True** function can be added to a rule to ensure names are returned in English.*

The iProperty data for a file is sorted into tabs along the top of the dialog box. The *iProperties* category of functions can be used to either read or set iProperty data in the tabs. The following describes how the iProperties in each of the tabs can be used in a rule.

Summary, Project, and Custom tabs

The syntax for setting or reading iProperty data from a document is similar for all the iProperties in the *Summary*, *Project*, and *Custom* tabs. The syntax for these include iProperties.Value at the beginning of the function followed by the component name (for assembly components), and the tab and iProperty names in brackets. For part files, the component name is not required.

```
iProperties.Value ("iProperty Tab Name", "iProperty Name")
iProperties.Value ("part:1", "iProperty Tab Name", "iProperty Name")
```

Some iProperty functions have snippets for use in a rule (e.g., Title, Part Number, Description, etc.). You can also explicitly enter a function to set or read other iProperties that do not have a specific snippet (e.g., Stock Number, Cost Center, etc.).

- In the following example, the Title and Part Number iProperties are populated based on the condition of an iLogic rule. The Title function is used in the rule to assign its value; however, the Stock Number function is typed manually as an explicit snippet did not exist.

```
If configuration = Large Then
iProperties.Value ("Summary", "Title") = Large Widget
iProperties.Value ("Project", "Stock Number") = LW-1621
ElseIf configuration = Regular Then
iProperties.Value ("Summary", "Title") = Regular Widget
iProperties.Value ("Project", "Part Number") = RW-1620
End If
```

- The following example shows how the value of the Author iProperty in the *Summary* tab can be read to drive another parameter value.

```
MyParameter = iProperties.Value ("Summary", "Author")
```

- The following examples show the syntax that should be used to generate a Custom iProperty. The second parameter in the brackets (e.g., **Number**, **Text**, **Date**, or **Boolean**) is the name of the custom iProperty. The syntax after the equal to (=) sign indicates the type of parameter. A number parameter's value is not included in brackets, a text parameter's value is included in brackets, a date has a custom syntax, and boolean parameters are entered as true or false.

```
iProperties.Value("Custom", "Number") = 16
iProperties.Value("Custom", "Text") = "Large"
iProperties.Value("Custom", "Date") = CDate("1/2/2013")
iProperties.Value("Custom", "Boolean") = true
```

Physical Tab

The *Physical* tab in the iProperties dialog box includes those properties that deal with the physical properties of the model. For example, mass, volume, area, and center of gravity.

To assign the material list as values in a parameter, use the following function: MutliValue.List("param") = iProperties.Materials.

- The *Material* iProperty can be used to set or read the material of a model. When setting a material using a rule, ensure that you use the exact material name from the active list.

```
iProperties.Material = "Steel"
iProperties.Material = "Steel, Mild"
```

Overwritten values are identified in the iProperty dialog box with the ⬟ *icon. To restore a calculated value, enter -1 as the value in the Rule.*

- The *Mass* and *Volume* iProperty values can be read or set using the **Mass** or **Volume** functions respectively. In either situation, if you write the mass or volume in a model, it overrides the calculated value. The first two functions below show the syntax for overwriting the mass and volume values in a part file. The second two show the syntax for overwriting the mass and volume values in a component of an assembly file. The part name or component name syntax in the brackets following the function can be used for either **Mass** or **Volume** functions.

```
iProperties.Mass = 0.187
iProperties.Volume = 5.158
iProperties.Mass("filename.ipt") = .0.187
iProperties.Volume("component name:1") = 5.158
```

Assigning a mass to a virtual component can be accomplished using the **iProperties.Volume ("component name:1")** syntax.

To read a Mass or Volume value, you set a parameter value equal to the function.

- The *Area* iProperty value can only be read from the model. The functions below show the syntax for reading the area value directly at the part level and for a component in an assembly.

```
parameter = iProperties.Area
parameter = iProperties.Area("component name:1")
```

- The *Center of Gravity* iProperty values can also only be read from the model. The lines of code shown below indicate the syntax for reading the X, Y, and Z directions of the center of gravity. The **pt** parameter is a special Autodesk Inventor API point parameter. The values for the center of gravity cannot be read using standard numeric parameters.

```
pt = iProperties.CenterofGravity
cx = pt.X
cy = pt.Y
cz = pt.Z
```

To read the center of gravity values for a component in a top-level assembly, you can use the following syntax.

```
pt = iProperties.CenterofGravity("component name:1")
cx = pt.X
cy = pt.Y
cz = pt.Z
```

- The **MaterialOfComponent**, **MassOfComponent**, **VolumeOfComponent**, and **AreaOfComponent** functions provide an alternative syntax to adding the component name in brackets at the end of a function to call out a component in a top-level assembly.

Part Color Property

The *Part Color* iProperty is not controlled in the iProperties dialog box. It is controlled using the Appearance controls in the Quick Access Toolbar or in the Appearance Browser. To read or write the Part Color with an iLogic rule, use the following syntax. The first two lines set the color to Green or that of the material. The third line reads the current color of the model to the **color_param** parameter.

```
iProperties.PartColor = "Green"
iProperties.PartColor = "As Material"
color_param = iProperties.PartColor
```

Material Properties

In addition to the *iProperties* category in the *System* tab, in the *Snippets* area, there is a *Material Properties* category. In this category, the snippets enable you to read the values of the model's assigned Material. For example, the snippets can be used to read the name of the material, its density value, its yield strength, etc. The syntax for all the Material Properties is similar. The material density and specific heat code is shown below.

```
materialDensity=ThisDoc.Document.ComponentDefinition.
Material.Density
specificHeat =ThisDoc.Document.ComponentDefinition.
Material.SpecificHeat
```

Practice 4a

Working with Conditional Statements

Practice Objectives

- Add a rule using an If...Then..ElseIf...Then... conditional statement.
- Create a rule that controls suppression of feature geometry based on a boolean parameter.
- Edit a rule to incorporate changes made to the design intent of the model.
- Edit a rule to assign colors to features based on the value of a user parameter.
- Develop an iLogic rule using a Select Case statement.

In this practice you will create more in-depth conditional statements in a rule by incorporating Else, Else If, and Case statements. The functions that are used in this practice will cover parameters, feature suppression, and feature color. These functions will be used to create variations in a model's design.

Task 1 - Add a rule using an If...Then..ElseIf...Then... conditional statement.

In this task, you add a rule to the model that will change the length of the model based on the value selected for a text based multi-value user parameter.

1. Open the model **brace.ipt**. The model displays as shown in Figure 4–12.

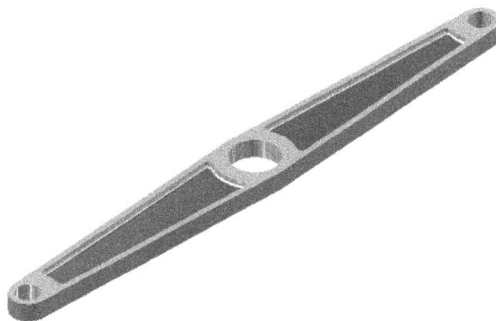

Figure 4–12

2. In the Quick Access Toolbar, click f_x (Parameters). The Parameters dialog box opens.

3. Sort the Parameter list using the **Renamed** option.

4. Review the two user parameters that have been created, as shown in Figure 4–13. The **size** parameter is a Multi-Value Text parameter and the **fillets** parameter is a True/False parameter. These parameters were created to provide for the model's design intent.

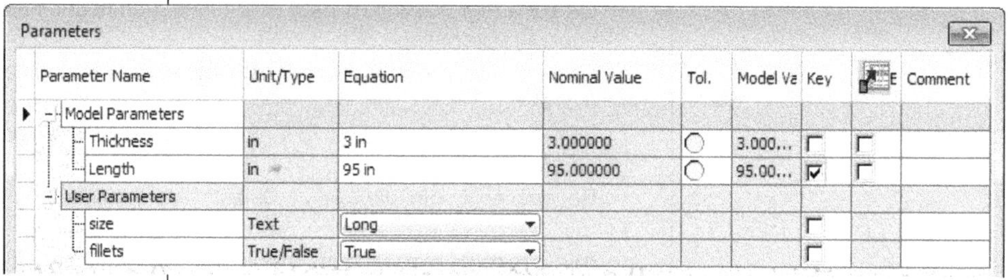

Parameter Name	Unit/Type	Equation	Nominal Value	Tol.	Model Va	Key	E	Comment
▶ − Model Parameters								
Thickness	in	3 in	3.000000	○	3.000...	☐	☐	
Length	in	95 in	95.000000	○	95.00...	☑	☐	
− User Parameters								
size	Text	Long				☐		
fillets	True/False	True				☐		

Figure 4–13

5. In the *Manage* tab>iLogic panel, click 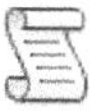 (Add Rule).

6. Enter **size** as the name of the rule. The Edit Rule dialog box opens.

7. In the bottom right area in the dialog box, click **If...Then...End If** to form the basis of the conditional statement.

8. Select the **User Parameters** node in the *Model* tab at the top in the Edit Rule dialog box. All user parameters are listed to the right, in the *Parameters* tab.

9. In the editor, select **My_Expression** and double-click on the **size** parameter that is listed in the *Parameters* tab to retrieve it into the rule. Alternatively, you can manually type the parameter name.

10. Enter **= "Short"** as the remainder of the conditional statement, as shown in Figure 4–14. Note the color coding of the statement.

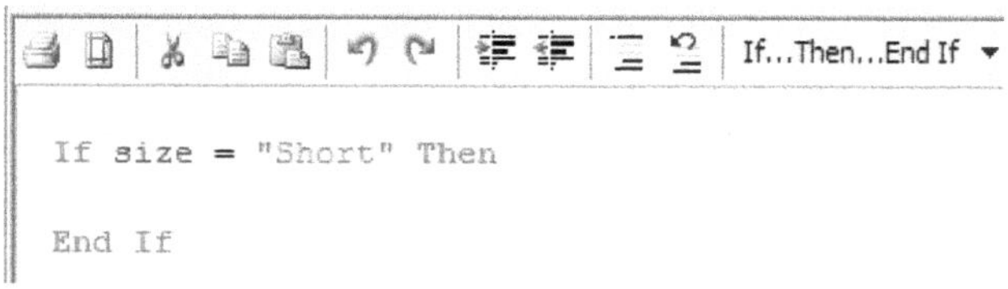

```
If size = "Short" Then

End If
```

Figure 4–14

11. Click **OK** to close the Edit Rule dialog box. Nothing changes in the model geometry. This is because the action for what to do if size = Short is not defined.

The next line in the rule must state what happens if the condition is met. In this case, if size = Short, the length of the model is to be made to equal 85.

12. Double-click on the **size** rule in the iLogic browser to edit the rule.

13. Position the cursor between the If and the End If lines.

As an alternative to selecting a model parameter, you can create a user parameter. Set the value of the model parameter to the user parameter's name. This makes locating the parameter easier when adding to a rule.

14. Select the **Model Parameters** node in the *Model* tab at the top in the Edit Rule dialog box. All model parameters are listed to the right, in the *Parameters* tab. Scroll through the list until you find the **Length** parameter. This parameter was renamed from its default d# expression. Double-click the **Length** parameter to add it to the rule. Alternatively, you can manually type the parameter name.

15. Complete the statement by typing **= 85**, as shown in Figure 4–15.

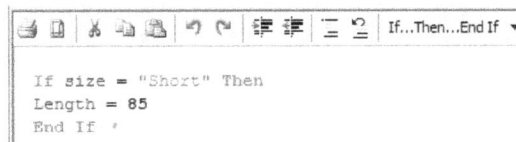

```
If size = "Short" Then
    Length = 85
End If
```

Figure 4–15

16. Click **OK** to close the Edit Rule dialog box.

17. Return to the Parameters dialog box and change the **size** parameter's value to **Short**. Note the change in the model geometry.

18. Change the **size** parameter's value to **Long**. Note that there is no change in the model. This is because the rule does not explain what to do for this condition.

19. Double-click on the **size** rule in the iLogic Browser to edit the rule.

20. Add a row before the End If statement by placing the cursor at the beginning of the End If line and pressing <Enter>.

21. In the If...Then...End If drop-down list, select **ElseIf...Then**.

22. Copy and paste the first condition and edit it. Add a second ElseIf...Then statement by copying the previous one. Edit the rule so that it displays as shown in Figure 4–16.

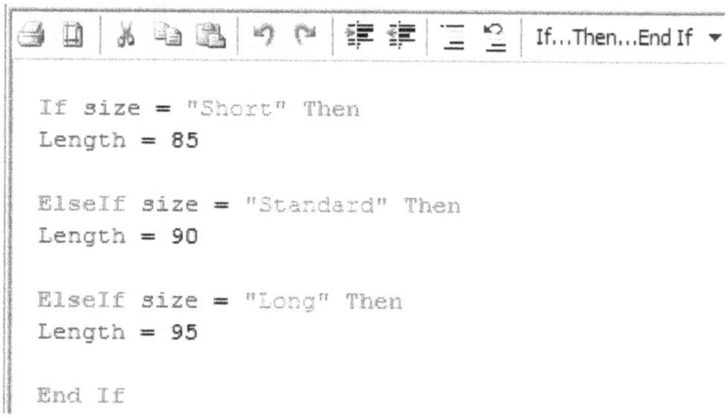

```
If size = "Short" Then
Length = 85

ElseIf size = "Standard" Then
Length = 90

ElseIf size = "Long" Then
Length = 95

End If
```

Figure 4–16

23. Close the Edit Rule dialog box.

24. Test the rule by changing the value of the **size** parameter. Verify that the Length dimension changes appropriately.

Task 2 - Create a rule that controls feature suppression based on a boolean parameter.

In this task you will use the **IsActive** Features function to control the suppression state of a fillet feature in the model. This is done by creating a conditional statement that verifies the value of a boolean parameter and based on this value, the suppression state is changed.

1. In the *Manage* tab>iLogic panel, click (Add Rule).

2. Enter **fillets** as the name of the rule. The Edit Rule dialog box opens.

3. In the bottom right area in the dialog box, click **If...Then...End If** to form the basis of the conditional statement.

4. Select the **User Parameters** node in the *Model* tab at the top in the Edit Rule dialog box. All user parameters are listed to the right in the *Parameters* tab.

5. In the editor, select **My_Expression** and double-click on the **fillets** parameter that is listed in the *Parameters* tab to retrieve it into the rule. Alternatively, you can manually type the parameter name.

6. Type **= "True"** as the remainder of the conditional statement.

7. Position the cursor between the If and the End If lines.

8. In the Edit Rule dialog box, in the *Snippets* area, expand *Features* and double-click on **IsActive** to add it to the rule.

9. In the *Model* tab at the top in the Edit Rule dialog box, select the **Fillet1** feature.

10. In the *Model* tab, near the right, select the *Names* tab. This lists the feature name of the selected feature.

11. In the **IsActive** function, select **featurename** and double-click on **Fillet1** in the *Names* tab to retrieve it into the rule. Alternatively, you can manually type the parameter name.

12. Type **= True** as the remainder of the statement.

13. Press <Enter> after the IsActive function to add a new line.

14. Expand the conditional statements drop-down list and select **Else**.

15. Copy and paste the **IsActive** function from the top portion of the rule and edit it as shown in Figure 4–17. A fillets = False condition is not required because false is implied as the only other option for the boolean parameter.

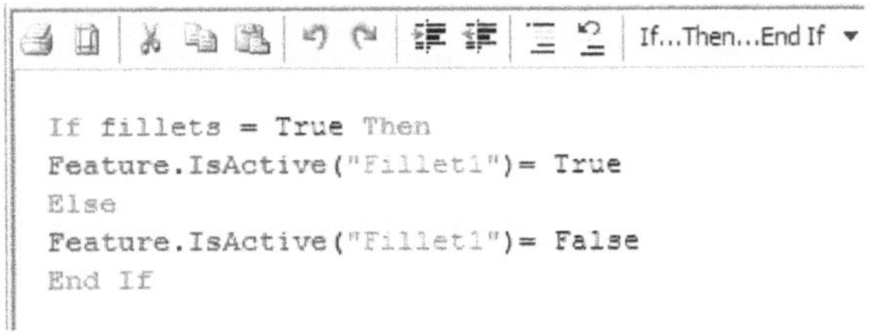

```
If fillets = True Then
Feature.IsActive("Fillet1")= True
Else
Feature.IsActive("Fillet1")= False
End If
```

Figure 4–17

16. Close the Edit Rule dialog box.

17. Test the rule by changing the value of the **fillets** parameter. Verify that the fillets are removed if the value for fillets = False and that they are displayed if the value is True, as shown in Figure 4–18.

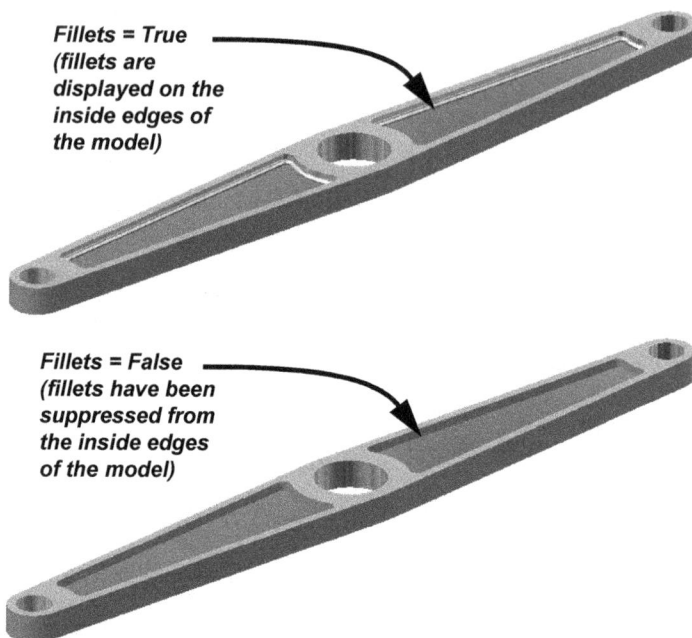

Fillets = True
(fillets are displayed on the inside edges of the model)

Fillets = False
(fillets have been suppressed from the inside edges of the model)

Figure 4–18

Task 3 - Edit the size rule and incorporate fillet control.

The current way that the rules have been completed requires the selection of one of the length values and entry for the boolean parameter for fillets. The design intent of the model has changed. The fillets are only required if the Standard or Long length values are selected. In the current scenario, this is required to be done manually. In this task, you will adjust the length rule to take this requirement into account.

1. Right-click the **fillets** rule and select **Suppress Rule**. You can also delete the rule. Suppressing can be used to prevent the rule from executing, while leaving it in the model in case the design intent changes and it is required again.

2. Double-click on the **size** rule in the iLogic browser to edit the rule.

3. Using the **IsActive** function in the *Features* snippets category, add the three new function lines to the code, as shown in Figure 4–19. To replace the "featurename" variable, use the *Model* tab to locate and select it for fillet1. This technique ensures that errors are not made in typing.

```
If size = "Short" Then
Length = 85
Feature.IsActive("Fillet1") = False

ElseIf size = "Standard" Then
Length = 90
Feature.IsActive("Fillet1") = True

ElseIf size = "Long" Then
Length = 95
Feature.IsActive("Fillet1") = True

End If
```

Figure 4–19

4. Close the Edit Rule dialog box.

5. Test the rule by changing the value for the **size** parameter. When the **Short** value is selected, the fillets are removed. When the **Standard** or **Long** values are selected, the fillets are included, as shown in Figure 4–20.

In the Short variation the fillets are suppressed

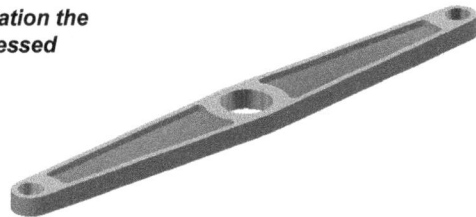

In the Standard variation the fillets are displayed

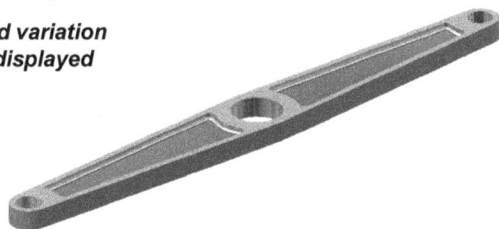

In the Long variation the fillets are displayed

Figure 4–20

Task 4 - Edit the size rule to assign colors to features based on the value of a user parameter.

The sizes of the three lengths only vary slightly and it is difficult to identify the variation by simply viewing the model. In this task, you will edit the size rule again and incorporate the use of the Color function to help identify the sizes.

1. Double-click on the size rule in the iLogic browser to edit the rule.

2. Using the **Color** function in the *Features* snippets category, add the three new function lines to the code, as shown in Figure 4–21. To replace the "featurename" variable, use the *Model* tab to locate and select the feature name for Extrusion4.

3. Enter the color variable that is assigned to the Extrusion4 feature at the end of each line, as shown in Figure 4–21. The color must be in quotes in order to identify it as a color in the Autodesk Inventor library.

```
If size = "Short" Then
Length = 85
Feature.IsActive("Fillet1") = False
Feature.Color("Extrusion4") = "Red"

ElseIf size = "Standard" Then
Length = 90
Feature.IsActive("Fillet1") = True
Feature.Color("Extrusion4") = "Blue"

ElseIf size = "Long" Then
Length = 95
Feature.IsActive("Fillet1") = True
Feature.Color("Extrusion4") = "Green"

End If
```

Figure 4–21

4. Test the rule by changing the **size** parameter, as shown in Figure 4–22.

In the Short variation, the fillets are suppressed and Extrusion4 is Red

In the Standard variation, the fillets are displayed and Extrusion4 is Blue

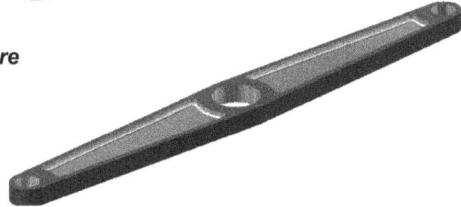

In the Long variation, the fillets are displayed and Extrusion4 is Green

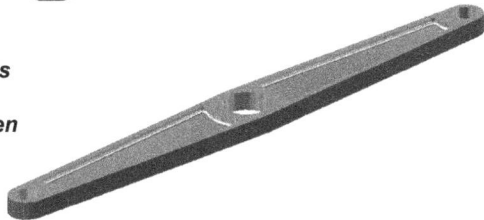

Figure 4–22

Task 5 - Convert the size rule to a Case statement.

Case statements can be used when multiple situations (or cases) are generated in a rule. It can be used in place of an extended block statement. In this task, the size rule will be edited using Case statement syntax to show an alternative to the extended block statement.

1. Edit the size rule.

2. Reusing existing code for the new Case statement eliminates having to retype and add functions. Select all the current code and click ⌐̲ (Comment).

3. Add some empty lines at the top of the rule.

4. With the cursor placed at the top of the rule, in the Keywords drop-down list, select **Select Case** to add the statement to the rule.

5. Enter or select the **size** parameter using the *Model* tab to add it after the Select Case statement, as shown in Figure 4–23. It displays in blue indicating that it is a parameter.

6. On the next line, in the Keywords drop-down list, select **Case**.

7. Place the cursor at the beginning of the Case statement and click ⊞ (Increase Indent). Indenting lines of code can be used to improve code readability.

8. Enter **"Short"** as the variable for the Case statement, as shown in Figure 4–23.

9. Copy the three function statements in the commented If statement for size = "Short" and paste them into the newly created Case statement for "Short".

10. Use ⊞ (Indent) and ⯑ (Uncomment), as required to obtain the Case statement shown in Figure 4–23.

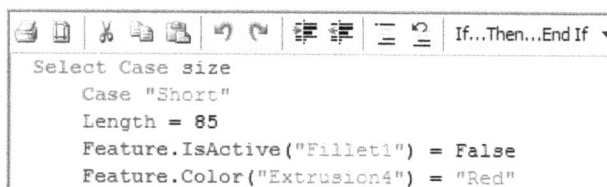

```
Select Case size
    Case "Short"
        Length = 85
        Feature.IsActive("Fillet1") = False
        Feature.Color("Extrusion4") = "Red"
```

Figure 4–23

11. Using any of the techniques discussed (manually typing code, adding snippets and editing it, or copy and pasting), complete the remaining two Case statements, as shown in Figure 4–24.

12. After the third Case statement, in the Keywords drop-down list, select **End Select** to add it to the rule as shown in Figure 4–24.

```
Select Case size
    Case "Short"
    Length = 85
    Feature.IsActive("Fillet1") = False
    Feature.Color("Extrusion4") = "Red"

    Case "Standard"
    Length = 90
    Feature.IsActive("Fillet1") = True
    Feature.Color("Extrusion4") = "Blue"

    Case "Long"
    Length = 95
    Feature.IsActive("Fillet1") = True
    Feature.Color("Extrusion4") = "Green"
End Select
```

Figure 4–24

13. Select and delete the commented **If...Else If** statements from the code.

14. Test the rule by changing the **size** parameter. The rule works in the same way as when the extended If statement was used. It is now written in a more readable format.

15. Save the model and close the window.

Practice 4b

Building a Logical Part Model II

Practice Objectives

- Add a rule to the model to control a feature's display based on the value entered for a user parameter.
- Edit a rule to include the use of the AND operator for defining multiple conditions that are to be met for a rule to execute.

Previously you prepared parameters for use with iLogic functionality. In this practice, you will add an iLogic rule to the model to control the following design criteria:

- Depending on the flange's number of ribs and outer diameter, the rib geometry automatically changes. The outer diameter is to be selected from a list of six options ranging from 50 mm to 100 mm. The number of ribs can be manually entered.

Task 1 - Add a rule to the model that controls the rib display.

In this task, you will add a rule to the model to suppress or unsuppress ribs based on the number of ribs in the model. Detailed steps have been omitted from the task; however, an image of the final rule is provided in Figure 4–26.

1. Open **iLogic_flange_2.ipt**. The model opens as shown in Figure 4–25. This is the model that was used in a previous practice when working with model and user parameters.

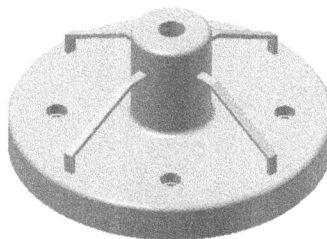

Figure 4–25

2. Open the Parameters dialog box and review the model and user parameters. These parameters are created and renamed to enable rule creation to meet a required design intent.

3. In the *Manage* tab>iLogic panel, click (Add Rule).

4. Enter **Rib_Type** as the name of the rule. The Edit Rule dialog box opens.

5. Click **If...Then...End If** to form the basis of the conditional rule.

6. Using the *Model* tab, populate the **My_Expression** placeholder with **number_of_ribs** model parameter. Edit the value that will be tested against such that it is < = 4.

7. The next line in the rule states what happens if the condition is met. In this case, if **number_of_ribs** < = 4, **Rib1** is to be displayed and **Rib2** is to be suppressed. Using the **IsActive** Feature function, add these actions to the rule.

8. The **Rib1_Fillet** and **Rib2_Fillet** features suppression state can be controlled based on the number of ribs. If Rib1 is displayed then **Rib1_Fillet** is displayed as well. If Rib2 is displayed then **Rib2_Fillet** is displayed as well. Incorporate these two actions into the rule.

9. Expand the conditional statements drop-down list and select **ElseIf... Then**.

10. Edit the **My_Expression** placeholder such that it tests for the condition of **number_of_ribs > 4**.

11. Copy and paste the four **IsActive** functions from the first condition in the rule and edit them so that **Rib2** and **Rib2_Fillet** are displayed.

The final Rule is shown in Figure 4–26.

```
If number_of_ribs <= 4 Then
Feature.IsActive("Rib1") = True
Feature.IsActive("Rib1_Fillet") = True
Feature.IsActive("Rib2") = False
Feature.IsActive("Rib2_Fillet") = False

ElseIf number_of_ribs > 4 Then
Feature.IsActive("Rib1") = False
Feature.IsActive("Rib1_Fillet") = False
Feature.IsActive("Rib2") = True
Feature.IsActive("Rib2_Fillet") = True

End If
```

Figure 4–26

12. Close the Edit Rule dialog box. If an error dialog box displays, review the code line that is causing the problem and make changes as required.

13. Test the rule by changing the **number_of_ribs** parameter to **6**. The model immediately updates and **Rib2** is now visible, as shown in Figure 4–27.

Figure 4–27

Task 2 - Modify the Rib_Type rule.

Changes are required to the **Rib_Type** rule to add a further condition controlling the two rib configurations. Along with the rib number, rib activation must also be based on the outer diameter.

1. Edit the **Rib_Type** rule.

2. Using **Copy** and **Paste**, duplicate the first condition and its associated functions.

3. Modify the conditions for the first two conditional statements to include an And operator. This is required to test the size of the flange's outer diameter at the same time as testing for the **number_of_ribs** parameter value. Modify the conditions as shown in Figure 4–28.

```
If number_of_ribs <= 4 And flange_OD <=60 Then
    Feature.IsActive("Rib1") = True
    Feature.IsActive("Rib1_Fillet") = True
    Feature.IsActive("Rib2") = False
    Feature.IsActive("Rib2_Fillet") = False

ElseIf number_of_ribs <= 4 And flange_OD > 60 Then
    Feature.IsActive("Rib1") = False
    Feature.IsActive("Rib1_Fillet") = False
    Feature.IsActive("Rib2") = True
    Feature.IsActive("Rib2_Fillet") = True

ElseIf number_of_ribs > 4 Then
    Feature.IsActive("Rib1") = False
    Feature.IsActive("Rib1_Fillet") = False
    Feature.IsActive("Rib2") = True
    Feature.IsActive("Rib2_Fillet") = True

End If
```

Figure 4–28

4. Close the Edit Rule dialog box.

5. Test the rule by changing the *flange_OD* to **80mm** and the **number_of_ribs** parameter to **4**. Does the model react as expected?

6. Change the **flange_OD** parameter value to **60 mm**. The model displays as shown in Figure 4–29. Does the model react as expected?

Figure 4–29

7. Save the model and close the window.

Practice 4c

Driving iProperties using iLogic Rules

Practice Objectives

- Develop an iLogic rule that uses a Select Case statement to control the color of a model.
- Develop an iLogic rule that drives how the *Description iProperty* field is filled out using the user parameter values.

In this practice, you will use iLogic to automate the process of assigning the color and description of a model. Both of these properties are controlled using iProperty functions. The model is shown in Figure 4–30. This model is intended to be used as a component in a top-level assembly.

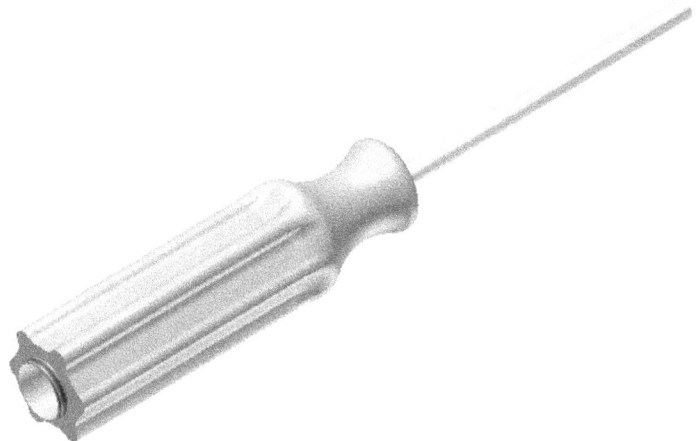

Figure 4–30

Task 1 - Open a model and review its existing parameters and iLogic rules.

In the first task you will open the model and review its current state. This model already contains custom parameters and two iLogic rules.

1. Open **body_screwdriver.ipt**.

2. Open the Parameters dialog box. Expand [icon] (Filter) and select **Key**. The Parameters dialog box is filtered to only display those parameters that have been identified as Key. In this model only the **EndType** and **Type** parameters are Key, as shown in Figure 4–31.

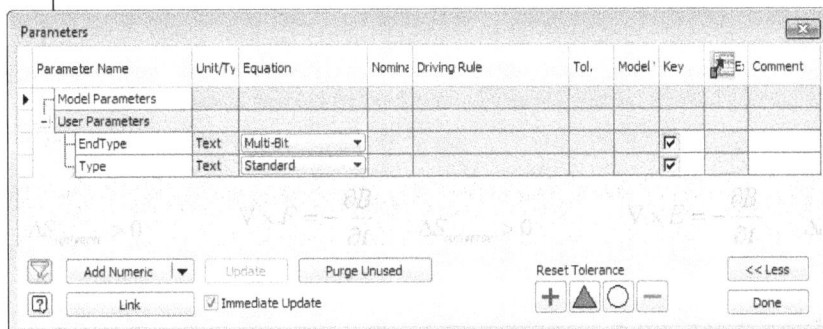

Parameter Name	Unit/Ty	Equation	Nomina	Driving Rule	Tol.	Model	Key		Comment
Model Parameters									
User Parameters									
EndType	Text	Multi-Bit					☑		
Type	Text	Standard					☑		

Add Numeric | ▼   Update   Purge Unused   Reset Tolerance   << Less

Link   ☑ Immediate Update   + ▲ ○ —   Done

Figure 4–31

3. Review the iLogic browser, as shown in Figure 4–32. The **EndCondition** and **Sizing** iLogic rules are created for you. Both of these rules use the **EndType** and **Type** parameters to control the geometry for the body of the screwdriver.

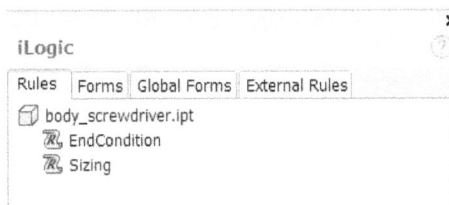

iLogic   ✕

Rules   Forms   Global Forms   External Rules

body_screwdriver.ipt
   EndCondition
   Sizing

Figure 4–32

Task 2 - Add an iLogic rule that adjusts the part color.

---

1. Open the Parameters dialog box, if not already open. The two Key parameters for this model are **EndType** and **Type**. Adjusting the **EndType** parameter controls whether the screwdriver body is solid with only one bit type (flat) or hollow with room to store a replaceable bit (Multi-Bit). This will be controlled in a top-level assembly. The **Type** parameter enables you to specify whether the screwdriver is a standard length or a smaller version. Select the different options for both parameters to manipulate the screwdriver body.

2. Click **Add Text** to add a text parameter. If the button is not displayed, open the drop-down list of the current parameter type option and select it from the list.

3. For the name of the parameter, type **Body_Color**.

4. Set this new parameter as **Key** by selecting the checkbox in the *Key* column.

5. Right-click in any of the cells for the **Body_Color** parameter's row and select **Make Multi-Value**. The Value List Editor dialog box opens.

6. In the top area in the dialog box, type **Red**, **Blue**, **Green**, **Yellow**, and **Black**, each on separate lines, as shown in Figure 4–33.

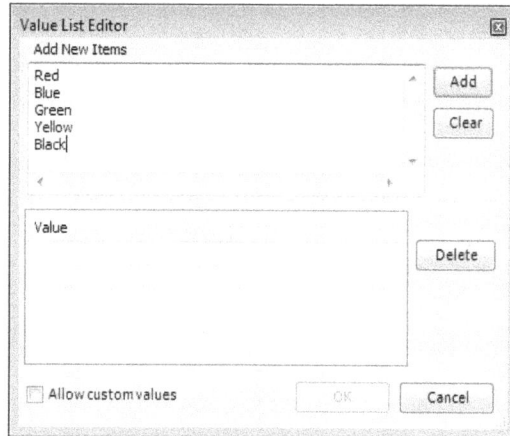

Figure 4–33

7. Click **Add**. The values are added to the lower area in the dialog box.

8. Click **OK**.

9. Close the Parameters dialog box. The color setting is not tied to the color property of the model, so you will not notice a change in color.

10. In the *Manage* tab>iLogic panel, click (Add Rule).

11. Type **Part_Color** as the name of the rule. The Edit Rule dialog box opens.

12. In the dialog box, in the *Rule Editor* area, expand the Keywords drop-down list and select **Select Case**. Enter **Body_Color** immediately after the Select Case statement, as shown in Figure 4–34.

13. Press <Enter> and enter **Case "Red"**.

*Using the Case structure (**Select Case**) is an alternative to using a series of If/Else statements to define the rule for assigning color to the model. It can be used to select from more than two selection options.*

14. Press <Enter> and expand the **iProperties** node in the *System* tab in the *Snippets* area. All snippets that deal with iProperties are listed. Double-click on **PartColor** to add the snippet to the rule.

15. At the end of the *iProperties.PartColor* snippet, enter **=** **"Red"**. The rule displays as shown in Figure 4–34.

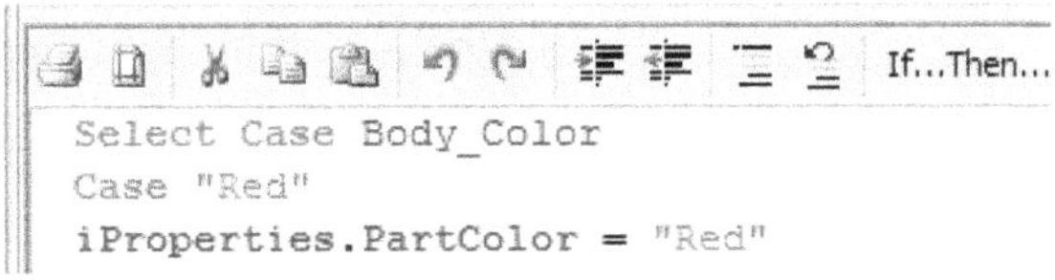

```
Select Case Body_Color
Case "Red"
iProperties.PartColor = "Red"
```

Figure 4–34

16. Copy and paste the last two lines in the rule an additional four times, and edit the rule, as shown in Figure 4–35, to include the other color options.

17. At the end of the rule, enter **End Select**, as shown in Figure 4–35. Alternatively, you can select the text from the Keywords drop-down list.

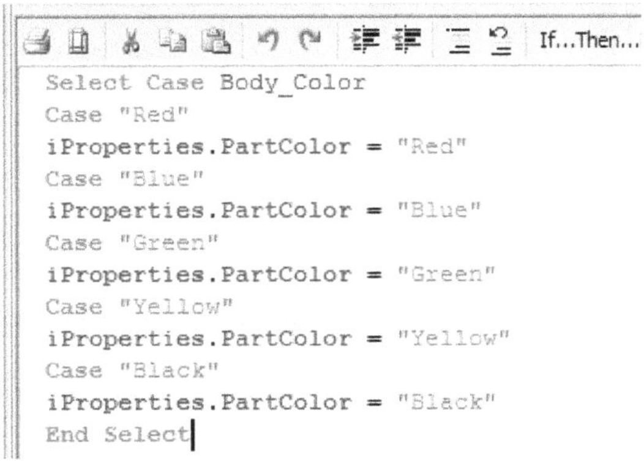

```
Select Case Body_Color
Case "Red"
iProperties.PartColor = "Red"
Case "Blue"
iProperties.PartColor = "Blue"
Case "Green"
iProperties.PartColor = "Green"
Case "Yellow"
iProperties.PartColor = "Yellow"
Case "Black"
iProperties.PartColor = "Black"
End Select
```

Figure 4–35

18. Click **OK** to complete the rule. The rule is listed at the bottom of the iLogic browser. Note that the model color updates to black. This was the last color entered in the multi-value list and was initially set as the default value.

19. Open the Parameters dialog box and set the **EndType** and **Type** parameters to **Multi-Bit** and **Standard** respectively, if not already set. Set the **Body_Color** parameter to **Green**.

20. Close the Parameters dialog box. The model color updates. The blade of the screwdriver remains gray because the feature color of this geometry was previously overridden (as gray) and is not affected by the part color change.

Task 3 - Add an iLogic rule that controls the Description iProperty value based on user parameter values.

1. Add another rule to the model.

2. Enter **Screwdriver_Description** as the name of the rule. The Edit Rule dialog box opens.

3. In the bottom right area in the dialog box, click **If...Then...End If** to form the basis of the conditional rule.

4. In the upper area in the dialog box, in the *Model* tab, select **User Parameters** for the model. All user parameters are listed to the right, in the *Parameters* tab.

5. Select **My_Expression** and double-click on the **Type** parameter that is listed in the *Parameters* tab to retrieve it into the rule. Alternatively, you can manually type the parameter name.

6. Enter **= "Standard"** as the remainder of the conditional statement, as shown in Figure 4–36.

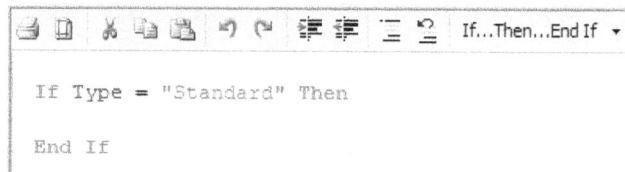

Figure 4–36

7. Press <Enter> at the end of the first line.

8. In the *Snippets* area, in the *System* tab, expand the **iProperties** node. Double-click on Description to add the snippet to the rule, as shown in Figure 4–37.

9. Press <Enter> at the end of the second line and enter **Else**, as shown in Figure 4–37.

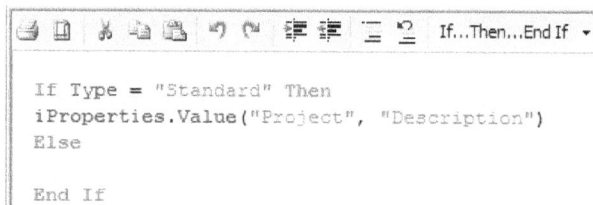

Figure 4–37

10. Place the cursor on a line under the Else statement.

11. Expand the **iProperties** node again and double-click on **Description** to add the snippet to the rule. Alternatively, you can also copy and paste line two.

12. Modify the end of each snippet to set the required description, as shown in Figure 4–38.

```
If Type = "Standard" Then
iProperties.Value("Project", "Description")= Body_Color & "-" & EndType
Else
iProperties.Value("Project", "Description")= Type & " " & Body_Color & "-" & EndType
End If
```

Figure 4–38

13. Click **OK** to complete the rule.

14. Right-click on the part name at the top of the Model browser and select **iProperties**.

15. Select the *Project* tab and verify the description displays the correct value, as shown in Figure 4–39.

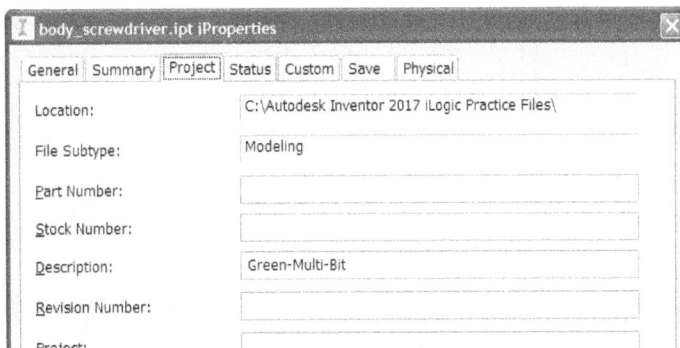

Figure 4–39

16. Close the iProperties dialog box.

17. Change the parameters in the model and verify that the *Description* iProperty updates as expected.

18. Save the model and close the file.

Chapter Review Questions

1. Which of the following Conditional Statements enable you to create a statement where if the condition is false the conditional statement is skipped entirely and no action is taken?

   a. If...Then...End If

   b. If...Then...ElseIf...Then...End If

   c. If...Then...Else...End If

2. The programming elements used to create a Case Structure type conditional statement can be used as an alternative to which of the following conditional statements?

   a. If...Then...End If

   b. If...Then...ElseIf...Then...End If

   c. If...Then...Else...End If

3. In an iLogic rule, which text color is used to identify the conditional statement elements?

   a. Black

   b. Blue

   c. Red

   d. Green

   e. Purple

4. The following is a valid conditional statement.

   ```
   If width < = 20 Then
   Else
   width = 20
   End If
   ```

   a. True

   b. False

5. Which of the following are valid operators that can be used in a conditional statement? (Select all that apply.)

   a. <

   b. >=

   c. ^

   d. And

   e. Or

6. Which of the following statements best describe the use of the following function?

    Parameter.Quiet = True

    a. Sets the boolean parameter called *Quiet* in the Parameters dialog box as **True**.

    b. Creates a boolean parameter called *Quiet* in the Parameters dialog box and sets its value as **True**.

    c. Prompts for an entry of a missing parameter value if encountered while running a rule.

    d. Prevents the display of an error message indicating that a parameter cannot be found in the model if a missing parameter is encountered while running a rule.

7. Which of the following statements best describes the following functions? (Select all that apply.)

    MultiValue.SetList("SIZE", 0.5, 0.75, 1.0, 1.25)
    Parameter.Param("Size").Comment = "List populated using the Material Library list"

    a. Creates a numeric parameter called **SIZE** in the Parameters dialog box.

    b. Sets the list of options for the multi-value parameter called **SIZE**.

    c. Sets the default value of the **SIZE** parameter to **0.5**.

    d. Sets the value of the *Comment* column for the **SIZE** parameter.

8. Which of the following Parameter functions reads the current value for the **d0** parameter in the model so that is can be used in a future function?

    a. values = MultiValue.List("d0")

    b. values = MultiValue.List("Part1:1", "d0")

    c. MultiValue.SetList("d0", 0.5, 0.75, 1.0, 1.25)

    d. MultiValue.List("d0") = GoExcel.CellValues ("filename.xls, "Sheet1", "A2", "A10")

9. In the following function, in what part file is the **d0** parameter located?

    Parameter (Make Path("ExtensionArm:1", Gasket:1"),"d0") = 22.7

    a. ExtensionArm

    b. Gasket

10. Which of the following can be controlled with a Feature function? (Select all that apply.)

    a. Feature Suppression

    b. Feature Color

    c. Model Color

    d. Thread Type (separate from the SetThreadAll function)

    e. Thread Designation

    f. Thread Class

11. Which of the following Physical iProperties can be set and read using iLogic snippets in the *iProperties* category? (Select all that apply.)

    a. *Material*

    b. *Mass*

    c. *Volume*

    d. *Area*

    e. *Center of Gravity*

12. The part color property is not controlled in the iProperties dialog box, therefore there is no iProperties Snippet that can control part color.

    a. True

    b. False

Command Summary

Button	Command	Location
NA	Thread	• *(double-click on an existing thread feature and select the Specification tab to access the Thread specifications defined in a rule)*

Assembly Rules and Functions

Just as part models can be automated, assembly models can also be automated using iLogic functions. Assembly functions enable you to control part level features from within the context of the top-level assembly, control the visibility of components, and control the constraints that position these components in the assembly.

Learning Objectives in this Chapter

- Describe the differences between the part model rule creation workflow and the elements that can be controlled using an assembly iLogic rule.
- Describe the importance and how to stabilize component names that are used in iLogic rules.
- Incorporate a Component function in an iLogic rule such that it either controls a component's suppression state or visibility or reads it to control the condition of an iLogic rule.
- Incorporate a Component function in an iLogic rule such that it controls a component's color.
- Incorporate a Component function in an iLogic rule such that it drives the replacement of assembly components with another component, an iPart, or a Level of Detail representation.
- Incorporate a Relationship function in an iLogic rule such that it either controls a component's constraint or joint suppression state or reads its state to control the condition of an iLogic rule.
- Incorporate a Relationship function in an iLogic rule such that it either controls an iMate suppression state or reads its state to control the condition of an iLogic rule.
- Create and place iLogic components in an assembly file.
- Modify iLogic component rules in subassemblies to ensure correct naming conventions are used when referencing subcomponents in a rule.

---

iLogic Workflow

Figure 5–1 shows the overall suggested workflow for the iLogic tools in the Autodesk® Inventor® software. The horizontal line at the top represents the high-level workflow and each of their sub-steps are detailed vertically below them. The highlighted column represents the content discussed in the current chapter.

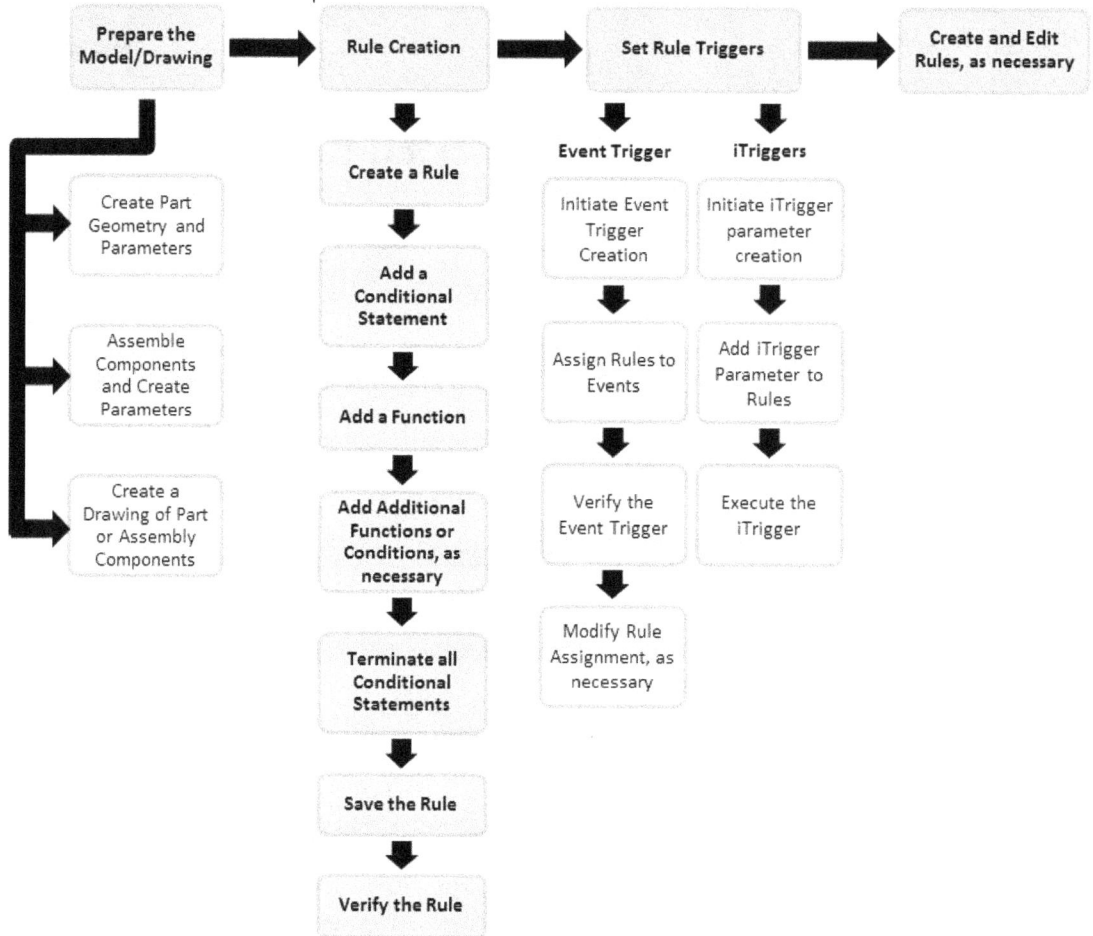

Prepare the Model/Drawing	Rule Creation	Set Rule Triggers	Create and Edit Rules, as necessary

Prepare the Model/Drawing
- Create Part Geometry and Parameters
- Assemble Components and Create Parameters
- Create a Drawing of Part or Assembly Components

Rule Creation
- Create a Rule
- Add a Conditional Statement
- Add a Function
- Add Additional Functions or Conditions, as necessary
- Terminate all Conditional Statements
- Save the Rule
- Verify the Rule

Event Trigger
- Initiate Event Trigger Creation
- Assign Rules to Events
- Verify the Event Trigger
- Modify Rule Assignment, as necessary

iTriggers
- Initiate iTrigger parameter creation
- Add iTrigger Parameter to Rules
- Execute the iTrigger

Figure 5–1

5.1 Assembly Rules

Previously, an overall workflow was provided to explain the process of working with iLogic, starting with preparing a model to successfully completing a rule. Here you understand how this workflow varies when iLogic is used to create logical assemblies.

The following can be accomplished using rules in an assembly model:

- Activate part and assembly features based on conditional statements.

- Activate assembly components based on conditional statements.

- Activate assembly constraints/joints based on conditional statements.

- Incorporate the execution of rules in assembly components from within a rule in the top-level assembly.

iLogic is VB-based, so visual basic code can be used.

In the example shown in Figure 5–2, iLogic is used to control the configuration of a screwdriver assembly. The rule that controls the geometry sets parameter values in the **screwdriver_body** part, and sets the **screwdriver_cap** and **screwdriver_bit** components as suppressed or visible along with their associated constraints. In addition, for two of the variations, the rule indicates the version of the **screwdriver_cap** iPart that is to be used.

```
If ScrewdriverType = "Standard Multi-Bit" Then
Parameter("screwdriver_body:1", "EndType") = "Multi-Bit"
Parameter("screwdriver_body:1", "Type") = "Standard"
Component.IsActive("Screwdriver_Cap") = True
Constraint.IsActive("CapBodyInsert") = True
iPart.ChangeRow("Screwdriver_Cap", "screwdriver_cap-01")
Component.IsActive("screwdriver_bit:1") = True
Constraint.IsActive("BitMate") = True
Constraint.IsActive("BitAngular") = True
Constraint.IsActive("BitCL") = True

ElseIf ScrewdriverType = "Standard Flat"  Then
Parameter("screwdriver_body:1", "EndType") = "Flat"
Parameter("screwdriver_body:1", "Type") = "Standard"
Component.IsActive("Screwdriver_Cap") = False
Constraint.IsActive("CapBodyInsert") = False
Component.IsActive("screwdriver_bit:1") = False
Constraint.IsActive("BitMate") = False
Constraint.IsActive("BitAngular") = False
Constraint.IsActive("BitCL") = False

ElseIf ScrewdriverType = "Stubby Flat"  Then
Parameter("screwdriver_body:1", "EndType") = "Flat"
Parameter("screwdriver_body:1", "Type") = "Stubby"
Component.IsActive("Screwdriver_Cap") = False
Constraint.IsActive("CapBodyInsert") = False
Component.IsActive("screwdriver_bit:1") = False
Constraint.IsActive("BitMate") = False
Constraint.IsActive("BitAngular") = False
Constraint.IsActive("BitCL") = False

ElseIf ScrewdriverType = "Stubby Multi-Bit" Then
Parameter("screwdriver_body:1", "EndType") = "Multi-Bit"
Parameter("screwdriver_body:1", "Type") = "Stubby"
Component.IsActive("Screwdriver_Cap") = True
Constraint.IsActive("CapBodyInsert") = True
iPart.ChangeRow("Screwdriver_Cap", "screwdriver_cap-03")
Component.IsActive("screwdriver_bit:1") = True
Constraint.IsActive("BitMate") = True
Constraint.IsActive("BitAngular") = True
Constraint.IsActive("BitCL") = True

End If
```

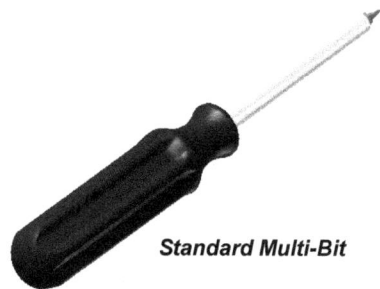

Standard Multi-Bit

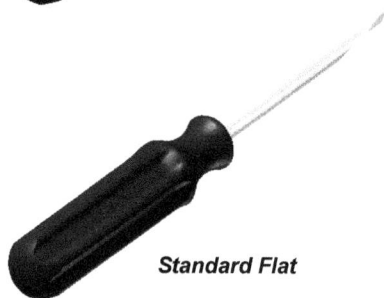

Standard Flat

Stubby Flat

Stubby Multi-Bit

Figure 5–2

As with most functionality in the Autodesk Inventor software, the use of iLogic also follows a workflow. This workflow for an assembly model can be broken down into four steps. Use the following general steps to create iLogic rules:

1. Prepare the model.
2. Rule creation.
3. Set rule triggers.
4. Create and edit rules, as necessary.

Figure 5–3 highlights the steps graphically. Additional in-depth information is included on each of these steps as you progress through this student guide.

Prepare the Model		Rule Creation		Set Rule Triggers		Create and Edit Rules, as necessary

Figure 5–3

Step 1 - Prepare the model.

When creating a top-level assembly that is going to incorporate iLogic rules, it is important to prepare the models:

- Verify that all part components contain the required dimensions, parameters, equations, and any required rules.

- Ensure that the feature dimensions capture the model's current geometric intent and have flexibility to change if updates are required.

This verification can be done by flexing the model which involves adjusting the key parameters to expected sizes to adjust the model correctly, and executing part level rules to ensure they react, as required.

As with creating any assembly, place and constrain the components to one another using standard Autodesk Inventor techniques. Verify that all components are fully constrained so that if constraints are used in the iLogic rule, the model reflects the design intent.

In addition, consider the following:

- Constraints or joints that are listed in the Model browser can be renamed to easily identify their intent in the assembly. To rename a constraint, select the constraint name once and type a new name.

- If any components in the assembly are iParts and if they are to be used in any iLogic rules, rename the iPart in the Model browser. As iParts are switched in an assembly, the name that displays in the Model browser typically updates to reflect the new component. If a rule drives the iPart, it must call out a static name in order to work. By renaming the **iPart** node in the Model browser with a new name, the iLogic rule has a static reference to modify.

Step 2 - Rule creation.

Rules can be created at a component or the top-level of the assembly model.

- To create rules in the components of an assembly, you either activate the component in the assembly or open each one of them independently. Rules created in components while in the context of an assembly are created in the same way as you would an individual part file.

- To create rules in the assembly, ensure that the top-level assembly is active and in the *Manage* tab>iLogic panel, click

  (Add Rule) to begin the creation of a new rule. The Edit Rule dialog box is similar to that used for part models, with the addition of the *File Tree* and *Files* tabs. These tabs provide alternate ways of listing the files in the assembly.

The steps for creating assembly rules are the same as those previously explained for creating rules in part models. The main difference is that assembly functions can be included in the assembly rule. Consider the following:

- In general, assembly rules should be used to perform assembly-level actions (e.g., change assembly parameter values, control assembly feature and component suppression states, set component alternates using iassemblies, set pattern quantity).

Detailed information on the Component functions are discussed later in the chapter.

Detailed information on the Constraint functions are discussed later in the chapter.

- Use the *Components* category in the *Snippets* area to include assembly component functions in the rule. This category enables you to control component activation, color, replacement, and visibility settings.

- Use the *Relationships* category in the *Snippets* area to include constraint functions in the rule. This category enables you to control the suppression state of assembly constraints and joints.

- Assembly and Part level rules can work together to achieve the required design intent. An assembly rule that should affect part level models must be written so that parameters are pushed down to the part level to ensure that part level rules are also run. This can be done using the following syntax:

    Parameter("Part:1", "param_name_part") = param_name_asm

- A Level of Detail Representation is required if an assembly rule is to control component suppression.

- Change any iPart model names listed in the Model browser. This ensures that any rules referencing the iPart have stable names and that the rule functions correctly as instances are changed in the assembly.

Once rules are created, you should verify that they are working as expected by flexing the model. Open the Parameters dialog box at the assembly level and modify values for any parameters that drive any of the rules. Ensure that values are tested for all scenarios to confirm that the assembly and the parts in the assembly behave as expected. Rules should also be explicitly run to verify it captures the design intent.

Step 3 - Set rule triggers.

Rule triggers enable you to define when a rule is triggered. The list of triggers vary slightly depending on whether a part, assembly, or drawing is active. Triggers are discussed later in this student guide.

Step 4 - Create and edit rules, as necessary.

Continue to add or edit rules, as required. A complete list of the rules can be reviewed in the iLogic browser. The order rules are listed in the iLogic browser can affect the resulting geometry. Drag and drop rules in the iLogic browser to verify that the order in which the rules are listed captures true design intent of the model.

If changes are required in a rule, right-click the rule name in the iLogic browser and select **Edit Rule**. Alternatively, you can also double-click the rule to open the Edit Rule dialog box.

5.2 Component Functions

The *Components* category in the *System* tab, in the *Snippets* area, provides access to a list of functions that are associated with component control in an assembly model. It can be used to set or read component activation, color, replacement, and visibility settings to create assembly configurations. The expanded list of Components functions is shown in Figure 5–4.

Hover the cursor over a function to display a short help line indicating its syntax and how it can be used in a rule.

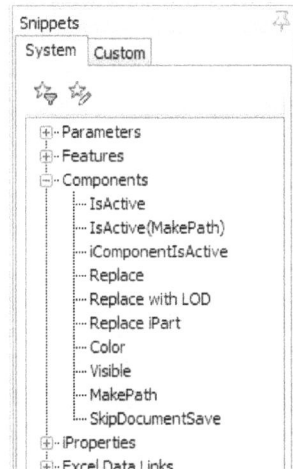

Figure 5–4

Component Activation Functions

There are three similar **IsActive** functions that can be used to either set or read the suppression state of a component (part or subassembly) in an assembly. Because of its ability to both set and read, it can be used as the test condition in an If or ElseIf statement or it can be used as a function to be executed if a condition is met. You can also incorporate both situations in a single rule.

- Component suppression should be controlled by an iLogic rule that exists at the same level as the component being controlled. For example, if a part component exists directly in the top-level assembly, the rule that controls its suppression should be at this level. To control a component in a subassembly, the **IsActive** function should be included in a rule in the subassembly and the parameter that is driving the rule can be included in another rule in the top-level assembly to drive the parameter.

- When using the **IsActive** functions in a top-level assembly, any rules containing these functions can only be executed in a Level of Detail (LOD) representation. If a rule that uses an **IsActive** function is executed in the Master LOD, an error message shown in Figure 5–5 is returned. A custom Level of Detail must be active to control suppression states using an iLogic rule.

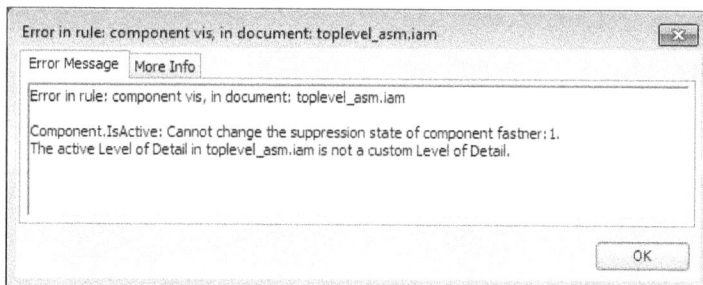

Error in rule: component vis, in document: toplevel_asm.iam	
Error Message	More Info

Error in rule: component vis, in document: toplevel_asm.iam

Component.IsActive: Cannot change the suppression state of component fastner:1.
The active Level of Detail in toplevel_asm.iam is not a custom Level of Detail.

OK

Figure 5–5

- When a component is suppressed using an **IsActive** function, its BOM Structure is set to Reference, as shown on the left in Figure 5–6. If the component is returned to an active state, it maintains its initial assigned BOM Structure (i.e., Normal, Inseparable, Purchased, or Phantom), as shown on the right in Figure 5–6.

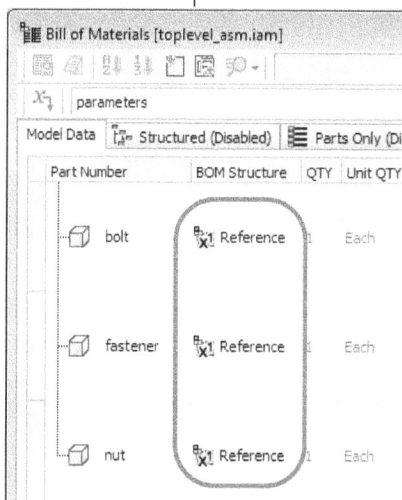

Bill of Materials [toplevel_asm.iam]			
parameters			
Model Data	Structured (Disabled)	Parts Only (Di	
Part Number	BOM Structure	QTY	Unit QTY
bolt	Reference	1	Each
fastener	Reference	1	Each
nut	Reference	1	Each

Bill of Materials [toplevel_asm.iam]			
Model Data	Structured (Disabled)	Parts Only (Di	
Part Number	BOM Structure	QTY	Unit QTY
bolt	Normal	1	Each
fastener	Normal	1	Each
nut	Normal	1	Each

The bolt, fastener, and nut components have been suppressed in a LOD configuration using the IsActive function. Their BOM Structure is set to Reference.

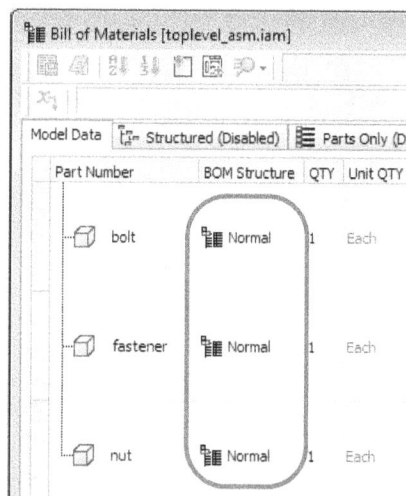

The bolt, fastener, and nut components BOM Structure is returned to Normal once they are displayed back in the assembly.

Figure 5–6

Examples of code using these functions are as follows:

- In the following example, the **IsActive** functions are activated based on the value of the **Fastener** parameter. Specifically, the Fastener, Nut, and Washer components are displayed if the value is **Yes** and they are suppressed if the value is **No**.

A value of 0 or 1 can also be used to specify the component state, where 0 is False and 1 is True.

```
If Fastener = "Yes" Then
Component.IsActive("Fastener:1")= True
Component.IsActive("Nut:1")= True
Component.IsActive("Bolt:1")= True
Else If Fastener = "No" Then
Component.IsActive("Fastener:1")= False
Component.IsActive("Nut:1")= False
Component.IsActive("Bolt:1")= False
End If
```

- In the following example, the **IsActive** function is used in both the conditional and action portions of a rule. The rule reads the suppression state of the Fastener component and if it is suppressed (False) the rule suppresses the Nut and Bolt. If displayed, the Nut and Bolt components are also displayed.

```
If Component.IsActive("Fastener:1")= False Then
Component.IsActive("Nut:1")= False
Component.IsActive("Bolt:1")= False
Else
Component.IsActive("Nut:1")= True
Component.IsActive("Bolt:1")= True
End If
```

- The **IsActive(MakePath)** function has the same purpose as the **IsActive** function, but it controls and reads suppression states for subassembly components. It provides an extra field that defines both the subassembly and part names.

```
Component.IsActive(MakePath("SubAssem1:1", "Part:1")
```

The **IsActive(MakePath)** function is best used to read information on a component's suppression state. It is recommended that you use the **IsActive** function in a subassembly to control its component suppression and the parameter driving the rule can be included in a top-level assembly rule to drive the parameter.

*If an iPart or iAssembly configuration is used in an assembly, it can be manually renamed to change the component name at which point the **IsActive** function can be used.*

- The **iComponentIsActive** function is another function that controls the suppression state. The main difference with this function is that it is specifically intended to read or set the suppression state for an iPart or iAssembly component. The syntax for this function is as follows, where the iPartA:1, is the specific name of the iPart configuration in the assembly.

    Component.iComponentIsActive("iPartA:1")

Visibility Functions

The **Visible** function can be used to either set or read the visibility of a component in the assembly. Because of its ability to both set and read, it can be used as a test condition in a conditional statement or it can be used as a function to be executed in a rule. The use of this function does not change the BOM structure or the iProperties of the assembly. It is simply for controlling component visibility. The syntax of the function is as follows:

    Component.Visible("ComponentName")

*To display a component for which the visibility has been toggled off in a rule, right-click in the Model browser and select **Visibility**. If the rule is run again it is removed from the display.*

- In the following example, the visibility of the Nut, and Washer have been set to **False**. When the rule is executed their visibility is cleared from the display in the assembly. These functions can be included in a conditional statement or on their own.

    ```
    Component.Visible("Fastener") = True
    Component.Visible("Nut") = False
    Component.Visible("Washer") = False
    ```

- The **Visible** function can also be used to read the visibility of a component. In the following example, the conditional statement involves reading the visibility state of a Fastener component. If it is displayed (i.e. True), the diameter of an assembly hole is set to **1**. If the Fastener visibility has been toggled off, the Hole is suppressed.

    ```
    If Component.Visible("Fastener") = True Then
    Hole_dia = 1
    Else If Component.Visible("Fastener") = False Then
    Feature.IsActive("Hole 1") = False
    End If
    ```

Color Functions

The **Color** function can be used to either set or read the color of a component in an assembly. In the following example, the color of the Base component changes to red if the value for the **Base_Color** parameter is set to **Red**, blue if set to **Blue**, and green if set to **Green**. The **Base_Color** parameter is created at the top-level assembly level as a multi-value parameter with **Red**, **Blue**, or **Green** as values.

```
If Base_Color = "Red" Then
Component.Color("Base:1") = "Red"
ElseIf Base_Color = "Blue" Then
Component.Color("Base:1") = "Blue"
ElseIf Base_Color = "Green" Then
Component.Color("Base:1") = "Green"
End If
```

Replacement Functions

There are three different replacement functions that can be used at the assembly level to drive component replacement in an iLogic rule. These include component replacement with another component, replacement with an iPart, or replacement with a Level of Detail representation. When replacing components with a function, the Autodesk Inventor software automatically searches in the following directories for the replacement model.

If the search paths do not describe the path of a replacing component, you can incorporate the relative path into the filename in the function.

- The same directory as the component being replaced.
- The directory of the assembly model.
- The *Workspace* folder of the current project.

Stabilizing Component Names

When working with assemblies a condition or function in an iLogic rule might contain specific reference to component names (i.e., replacement). If an assembly component's name changes either due to the execution of the rule or due to an explicit replacement, any rules that reference the initial component name fails. To avoid this situation, when working with assemblies, it is important to *stabilize* component names.

When components are assembled, the Model browser is populated, by default, using the Part Name that displays in the directory from which it is assembled. If the component is replaced with another component or as an instance from a factory table, the Part Name updates reflecting the change. To stabilize the component's name you can edit the name in the Model browser to a custom occurrence name. Once a component has been stabilized using a custom occurrence name, the *Name* field in the *Occurrence* tab in the iProperties dialog box updates with the stabilized name, as shown in Figure 5–7.

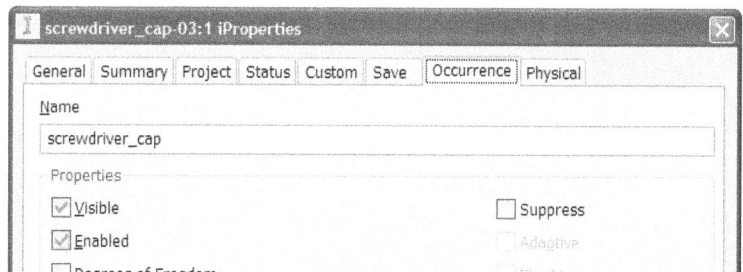

Figure 5–7

This custom occurrence name does not change regardless of component replacement. The new name should be the name referenced in all iLogic rules.

Replacing components with other components

The **Replace** function can be used to replace a component with another. The function can be executed based on the outcome of a conditional statement or can be used as an explicit statement in a rule. The syntax of the function is as follows:

Component.Replace("ComponentToReplaceName", "OtherPartName", <ReplaceAll>)

Where:

"ComponentToReplaceName" represents the name of the component (part or subassembly) being replaced.

"OtherPartName" represents the component (part or subassembly) to be used as the replacement component.

If you want the occurrence name to be the same as the initial component, rename it to a temporary name and then rename it back. The process of renaming stabilizes the component, it is not dependent on the name used.

Before using any of the replacement functions, consider renaming component names in the Model browser to lock their naming scheme even after replacement.

<ReplaceAll> is a boolean value that determines whether the replacement affects all instances of the component or just a single instance. Set to **True** to replace all instances and **False** to replace only the single named instance.

- In the following example, the value of the **Usage** parameter determines which model (**Base_Low**, **Base_Med**, or **Base_High**) is used in the assembly. The True boolean value at the end of the function ensures that all instances of Base are replaced with the specified part model.

```
If Usage = "Low" Then
Component.Replace("Base", "Base_Low.ipt", True)
ElseIf Usage = "Medium" Then
Component.Replace(Base", "Base_Med.ipt", True)
ElseIf Usage = "High" Then
Component.Replace(Base", "Base_High.ipt", True)
End If
```

Replacing components with iPart instances

The **Replace iPart** function should be used as an alternative to the **Replace** function when the component being used as the replacement component is an iPart. The syntax of the function is as follows:

```
Component.ReplaceiPart("ComponentToReplaceName",
"OtherPartfilename", <ReplaceAll>), rowNumber)
```

Where all of the fields are equivalent to those for the Replace function, but the *rowNumber* field is included to specify the specific iPart instance that is used in the assembly.

Once a component is replaced with an iPart model, you can use ***iPart.ChangeRow*** *or* ***iPart.FindRow*** *to change the specific configuration in a rule. The replacement rule that changed the model cannot execute because the existing component name in the model has now changed.*

- In the following example, the value of the **Usage** parameter determines which instance of the PartA component is replaced in the model. Row 1, 2, and 3 in the iPart table for the PartA component has been defined with values that pertain to Low, Medium, and High use materials.

```
If Usage = "Low" Then
Component.ReplaceiPart("Base", "PartA.ipt", True, 1)
ElseIf Usage = "Medium" Then
Component.ReplaceiPart("Base", "PartA.ipt", True, 2)
ElseIf Usage = "High" Then
Component.ReplaceiPart("Base", "PartA.ipt", True, 3)
End If
```

- In the following example, an extra field is added after the *rowNumber* field to assign a value for a custom parameter. In this case, the custom parameter is a diameter value that is set to be variable in the PartA component. When the instance in row 1 is used, the diameter of the hole is set at 20, for row 2 it is 15, and for row 3 it is 10.

```
If Usage = "Low" Then
Component.ReplaceiPart("Base", "PartA.ipt", True, 1, 20)
ElseIf Usage = "Medium" Then
Component.ReplaceiPart("Base", "PartA.ipt", True, 2, 15)
ElseIf Usage = "High" Then
Component.ReplaceiPart("Base", "PartA.ipt", True, 3, 10)
End If
```

Replacing components with level of detail reps

*The **Replace with LOD** function is applicable for subassembly replacement only, as part components do not have LOD representations.*

The **Replace with LOD** function should be used as an alternative to the **Replace** function when a specific level of detail representation in a subassembly is being used as the replacement component in another subassembly. It can be used to replace a subassembly with a Level of Detail representation in a new assembly or to switch between Level of Detail representations in the same assembly. The syntax of the function is as follows:

```
Component.Replace("AsmtoReplaceName",
"OtherAsmName<LevelOfDetail>", <ReplaceAll>)
```

Where:

"AsmtoReplaceName" represents the name of the subassembly component being replaced.

"OtherAsmName<LevelofDetail>" represents the subassembly component and the name of the Level of Detail representation that is to be used as the replacement component.

<ReplaceAll> is a boolean value that determines whether the replacement affects all instances of the component or just a single instance. Set to True to replace all instances and False to replace only the single named instance.

*Unlike replacing iPart instances where you must use the iPart.ChangeRow or iPart.FindRow functions to change the specific configuration in a rule, the **Component. Replace** function can be used to replace level of detail representations in a single assembly.*

- In the following example, the value of the **Detail** parameter determines which Level of Detail representation of the Piston subassembly is replaced in the model. When the **Detail** parameter is **Low**, the Low LOD representation for the Piston assembly is replaced, if the **Detail** parameter value is **All**, the Master LOD representation for the Piston assembly is used.

```
If Detail = "Low" Then
Component.Replace("Piston:1", "Piston.iam<Low>, True)
ElseIf Detail = "All" Then
Component.Replace("Piston:1", "Piston.iam<Master>, True)
End If
```

MakePath

The **MakePath** function in the *Components* category is used to specify the path to a subassembly component. This is commonly used if multiple components with the same name exist in the top-level assembly. The MakePath function can be used to replace the component name field in any functions. For example, when the Color function is added to a rule it displays as shown in the top line of the following example. By selecting **"PartA:1"** and replacing it with the **MakePath** function, the new line displays as shown in the second line.

```
Component.Color("PartA:1")
Component.Color(MakePath("SubAssem1:1", "Part2:1"))
```

5.3 Relationship Functions

The *Relationships* category in the *System* tab in the *Snippets* area provides access to functions that control assembly constraint and joint suppression. These functions can be used to create assembly level configurations that vary component placement. The expanded list of Relationship functions is shown in Figure 5–8.

Hover the cursor over a function to display a short help line indicating its syntax and how it can be used in a rule.

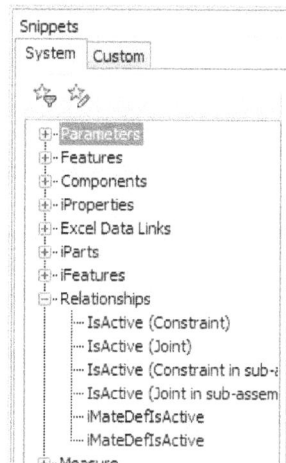

Figure 5–8

When preparing your model to use Relationship functions, consider the following:

- Constraints and Joints should be renamed to help describe their purpose in the assembly. This helps in selecting and using the correct constraint names in an iLogic rule.

- In an Autodesk Inventor assembly model, multiple constraints and joints cannot exist that are in conflict with one another. For example, two mate constraints that constrain the component in the XY plane cannot exist in the model at the same time. To workaround this design limitation, create the initial constraint or joint and then manually suppress it and create a second one. Once you have all constraint/joint variations (suppressed or not) you can begin rule creation.

The **Relationships** functions enable you to control constraints, joints, and iMate definitions in a top-level or subassembly model.

Constraint & Joint Suppression

There are four **IsActive** functions available to control the suppression state of a constraint or joint. The difference between these functions is that two are used for constraints and two are used for joints. One set is used for the top-level assembly constraints/joints (**IsActive**) and the other is used to control constraints/joints in subassemblies (**IsActive (... in sub-assembly)**). The function used for sub-assembly constraints and joints includes the path to locate the constraint/joint in the assembly.

Similar to many functions, the **IsActive** function has the ability to both set and read states in the assembly. For example, it can be used as the test condition in an If or ElseIf statement or it can be used as a function to be executed if a condition is met. You can also incorporate both situations in a single rule.

- In the following example, the **IsActive (Constraint)** functions are activated based on the value of the **Configuration** parameter. Specifically, the **Mate:1** constraint is active if the **Full_Movement** value is assigned for *Configuration*. If the value is set to **No_Movement**, the **Mate:1** constraint is active.

  ```
  If Configuration = "Full_Movement" Then
  Constraint.IsActive("Mate:1")= False
  Else If Configuration = "No_Movement"Then
  Constraint.IsActive("Mate:1")= True
  End If
  ```

A value of 0 or 1 can also be used to specify the constraint's state, where 0 is False and 1 is True.

- In the following example, the **IsActive (Joint)** functions are activated based on the value of the **Configuration** parameter. Specifically, the **Ball** joint is active if the **Full_Movement** value is assigned for *Configuration* and the **Rigid** joint must be suppressed. If the value is set to **No_Movement**, the **Rigid** joint is active and the **Ball** joint is suppressed. In this situation you must add both joints to the assembly. To do so, create the first, suppress it, and then add the second.

  ```
  If Configuration = "Full_Movement" Then
  Joint.IsActive("Ball:1")= True
  Joint.IsActive("Rigid:1")= False
  Else If Configuration = "No_Movement"Then
  Joint.IsActive("Ball:1")= False
  Joint.IsActive("Rigid:1")= True
  End If
  ```

- In the following example, the **IsActive (Constraint)** function is used in the condition of a rule. The rule reads the suppression state of the **Mate:1** constraint. If it is active, the rule sets the **Screw_Sub:1** color to **Red**, visually indicating that the component is fully constrained. If the constraint is suppressed, the color turns green indicating it can be rotated. **IsActive (Joint)** functions can also be used in the condition of a rule.

```
If Constraint.IsActive("Mate:1") = "True" Then
Component.Color("Screw_Sub:1") = "Red"
ElseIf Constraint.IsActive("Mate:1") = "False" Then
Component.Color("Screw_Sub:1") = "Green"
End If
```

- The **IsActive (Constraint in sub-assembly)** and **IsActive (Joint in sub-assembly)** functions are used in a similar way to the IsActive functions. However, they include syntax that points to the constraint or joint used to locate components in a subassembly versus a top-level assembly constraint or joint.

iMate Definition Suppression

There are two *Relationships* functions that can be used to read or control iMate suppression in an assembly model. The difference between the two is that one is used for iMates that exist in the top-level assembly model and the other is for iMates in subassemblies. Similar to controlling constraint and joint suppression, the difference is in the syntax for locating the iMate. For sub-assembly level iMates, you must include the assembly name in which the iMate exists. Both functions have the same name in the Snippets list, **iMateDefIsActive**. To distinguish between the two, hover the cursor over the function to review its syntax before adding it to a rule.

*The **iMateDefIsActive** function can also be used as a condition in a rule.*

- In the following example, the **iMateDefIsActive** function is used in a rule to control the suppression state of the iMate called **iMate_mount_holes**. Based on the value of the **Hole** parameter, the suppression state of an assembly level hole feature and its associated iMate is suppressed or not.

```
If Hole = "No" Then
Feature.IsActive("Mount_Hole") = False
Constraint.iMateDefIsActive("base:1", "iMate_mount_holes") = False
ElseIf Hole = "Yes" Then
Feature.IsActive("Mount_Hole") = True
Constraint.iMateDefIsActive("base:1", "iMate_mount_holes") = True
End If
```

5.4 iLogic Components in Inventor Assemblies

Components are assembled to one another to create an assembly model. When working with part or assembly components that contain iLogic rules, it is possible to assemble it in one of the following two ways:

- Assemble the actual source component model that was created in the Autodesk Inventor software.

- Assemble an iLogic component of the source model.

In the case of an assembly that is placed as an iLogic component, its component models are also copied.

In the second scenario, the iLogic component is created as a copy of the source model. During placement, you are prompted to enter values for the Key parameters. If no parameters are marked as **Key**, the entire list is displayed. Any iLogic rules are also copied to the iLogic component and they can be executed once in the context of the assembly. Using an iLogic component model in an assembly enables you to make changes to the component's geometry without affecting the source component model, and vice-versa.

General Steps

Use the following general steps to create an assembly and insert iLogic components:

1. Create an assembly file and save it.
2. Add and constrain iLogic components in an assembly file.
3. Repeat Step 2 to add additional components.
4. Modify iLogic components.
5. Save the assembly file.

Step 1 - Create an assembly file and save it.

You can also open the New File dialog box by clicking ▢ (New) in the Quick Access Toolbar or in the Application Menu.

Similar to creating a standard assembly file, the first step is to start a new file based on the assembly template. This can be done using ▢ (New) in the *Get Started* tab>Launch panel or using 🔧 in the *My Home* tab.

- The assembly must be saved before continuing. Save the assembly with a descriptive, unique name.

Step 2 - Add and constrain iLogic components in an assembly file.

In the *Assemble* tab>Component panel, expand the Place

options and click ![icon] (Place iLogic Component) to add a component as an iLogic component in the assembly. In the Place iLogic Component dialog box, select the component to add and click **Open**. The Place iLogic Component dialog box displays similar to that shown in Figure 5–9.

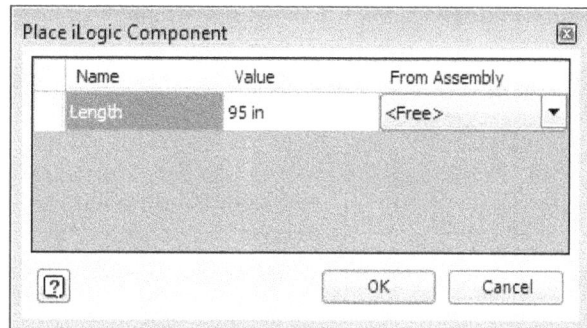

Place iLogic Component			
Name	Value	From Assembly	
Length	95 in	<Free>	▼

| ? | | OK | Cancel |

Figure 5–9

The columns in the Place iLogic Component dialog box are described in the following table.

Column Name	Description
Name	Lists all parameters (model and user) in the source component that is being assembled. If any parameter was identified as Key in the source file's Parameters dialog box, only that parameter is displayed in the *Name* column.
Value	Enables you to edit parameter values in the model before component placement.
From Assembly	Enables you to assign the model parameter to be controlled by an assembly parameter. All assembly parameters must be created before assembling the iLogic component. Once created, it is listed in the From Assembly drop-down list and can be assigned to control the component parameter.

Enter new values and reference top-level assembly parameters, as required. Click **Open** to create the iLogic component and place it in the assembly.

Naming Convention

It is good practice to ensure that the iLogic source model is stored in another directory, separate from the iLogic component and the assembly.

When an iLogic component is assembled, a copy of the source model is generated and stored in the same directory as the assembly that it is being assembled into.

The naming convention for iLogic components is such that a -01 is appended to the component name for each component that is placed in the assembly. Multiple instances of the same component reference the same copied file.

Consider the following when creating a top-level assembly:

- Assembly components can be a mix of standard components and iLogic components.

- iLogic components should be fully constrained in the assembly. All of the standard assembly tools and constraints can be used on iLogic components.

- iLogic components can be replaced with their source model using the **Component>Replace** functionality.

Step 3 - Repeat Step 2 to add additional components.

Continue to add iLogic or standard components to the assembly, as required.

Step 4 - Modify iLogic components.

You should review any existing iLogic rules in the assembly models (parts or assemblies) to verify if component names are referenced in any rules. When copied, an iLogic component and any sub-components are renamed. If rules exist that refer to a source file naming convention the rules do not execute correctly. iLogic rules are copied with iLogic components; however, the names of components used in these rules do not update automatically to incorporate the addition of -01 that is appended to the end of iLogic components. Edit and review the rules in each component, as required, to ensure that the naming conventions are appropriate for the assembly.

- As an alternative, you can also edit the naming convention in the source model's rules to append the -01 characters, prior to assembling and creating the iLogic component. If renamed prior to assembling ensure that you edit the iLogic rules back to their correct naming convention after the iLogic component is created.

Step 5 - Save the assembly file.

Save the assembly file once all components have been assembled.

Practice 5a

Building a Logical Assembly Model

Practice Objectives

- Verify part level iLogic rules before referencing them in top-level assembly rules.
- Add multi-value parameters to the top-level assembly file.
- Automate the process of selecting variations of an assembly model using iLogic rules in the top-level assembly.
- Add comments to a rule to provide information on the purpose of the code.

In this practice you will use iLogic to automate the process of selecting the size, color, and type variations of a screwdriver. The initial model is shown in Figure 5–10.

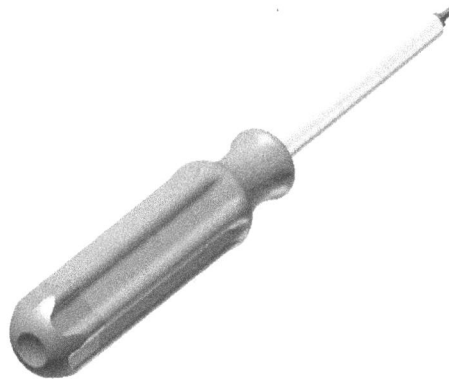

Figure 5–10

Task 1 - Open a screwdriver assembly and review the model.

In the first task you will open the assembly and review its current state. This assembly contains one iPart and two other parts that already contain iLogic code.

1. Open **iLogic_Screwdriver.iam** from the *Screwdriver* folder. Note that the screwdriver cap has a dimple in it.

2. In the Model browser, expand **screwdriver_cap-03.ipt**, if not already expanded.

3. Right-click on Table and select **Change Component**, as shown in Figure 5–11.

Figure 5–11

4. In the *Keys* tab, select **Extra Long Cap** and select **Standard Cap** from the drop-down list. (Alternatively, select the *Tree* tab and select **Standard Cap**.) Click **OK**. Note that the cap gets shorter and the dimple at the end of the cap is removed.

5. Double-click on **screwdriver_bit** to activate the component.

6. Open the Parameters dialog box.

The selected filter setting is maintained until explicitly changed.

7. In the Parameters dialog box, expand ⬚ (Filter) and select **Key**, if not already set. The Parameters dialog box displays as shown in Figure 5–12. Note that only one parameter (**BitType**) has been identified as a **Key** parameter in the model.

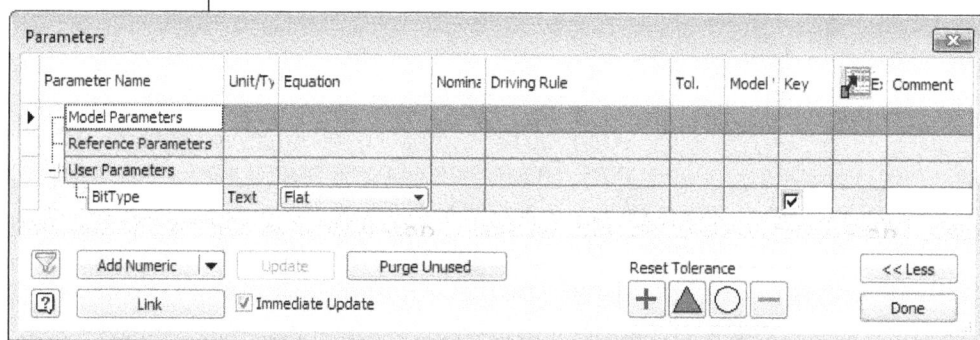

Figure 5–12

8. Enable **Immediate Update** in the Parameters dialog box, if not already enabled. This ensures that the parameter updates immediately when changed, instead of only when the dialog box is closed.

9. Adjust the **BitType** parameter value and note that the component geometry adjusts accordingly. Figure 5–13 shows the four bit types that are available.

Figure 5–13

10. Set the **BitType** parameter to **Flat** and close the Parameters dialog box.

11. In the iLogic browser, right-click on the **BitType_Code** rule and select **Edit Rule**. The Edit Rule dialog box opens as shown in Figure 5–14. This rule controls the change in geometry required to create the bits.

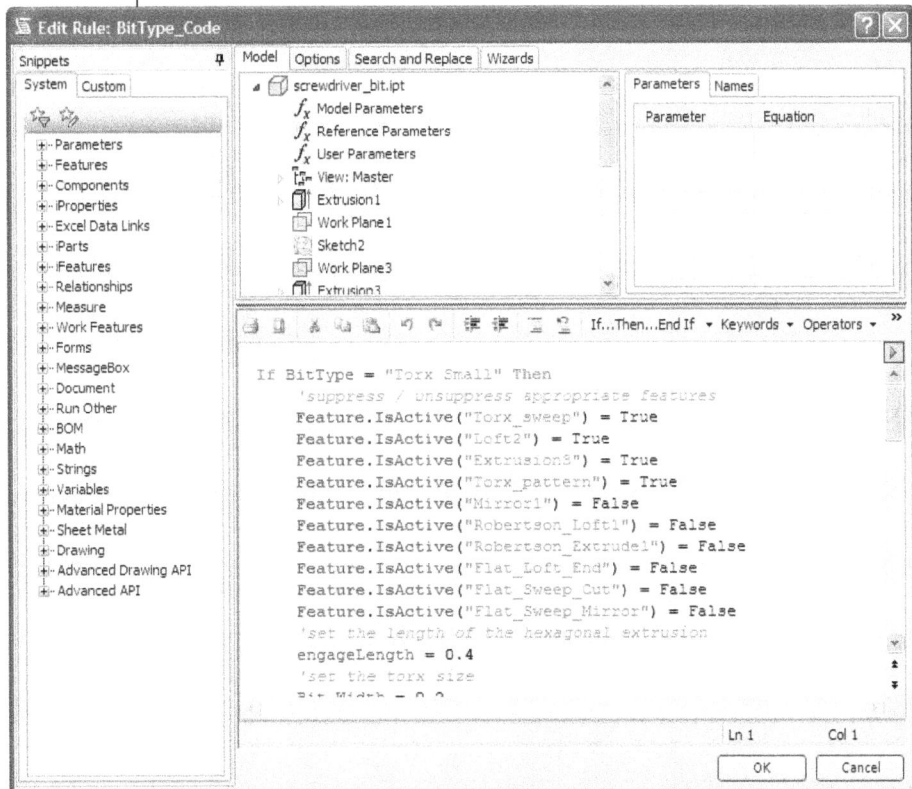

Figure 5–14

Any text that is preceeded with a single quote (') is commented text and displays as gray in the rule. A commented text line helps to describe the purpose of the rule or code.

The blade of the screwdriver remains gray because the feature color of this geometry is overridden (as gray) and is not affected by the part color change.

12. Review the code controlling the bit type. For example, in the first If statement, if the **BitType** parameter is set to **Torx Small**, the features of the model are all either set to **True** (unsuppressed) or **False** (suppressed), and the sizes are varied to provide the required geometry. Additional ElseIf statements are written to define the other variations for the BitType.

13. Close the Edit Rule dialog box.

14. Activate the top-level assembly.

15. Double-click on **screwdriver_body** to activate the component. The model has four rules: **EndCondition**, **Sizing**, **Part_Color**, and **Screwdriver_Description**. The **Part_Color** and **Screwdriver_Description** rules are the same as those rules you created earlier in this student guide. They control the model color and description iProperties.

16. Open the Parameters dialog box and change the **Body_Color** parameter value to **Red**. Close the dialog box.

17. Right-click on **screwdriver_body** in the Model browser and select **iProperties**. Select the *Project* tab. Note that the *Description* field is populated with the correct information (Red-Multi-Bit). This was driven based on the **Screwdriver_Description** rule.

18. Return to the assembly and update it, if required.

Task 2 - Add iLogic parameters to the assembly.

---

1. In the top-level assembly, open the Parameters dialog box and add a new text parameter called **ScrewdriverType**. Set it as a **Key** parameter.

2. Set the new parameter as a multi-value parameter and add the following values: **Standard Multi-Bit**, **Standard Flat**, **Stubby Multi-Bit**, and **Stubby Flat**, as shown in Figure 5–15.

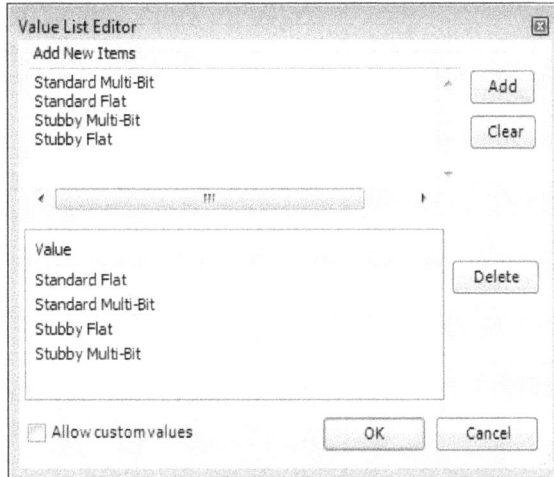

Figure 5–15

3. Add a new text parameter called **ScrewdriverColor**. Set it as a **Key** parameter.

4. Set the new parameter as a multi-value parameter and add the following values: **Red**, **Blue**, **Green**, **Yellow**, and **Black**, as shown in Figure 5–16.

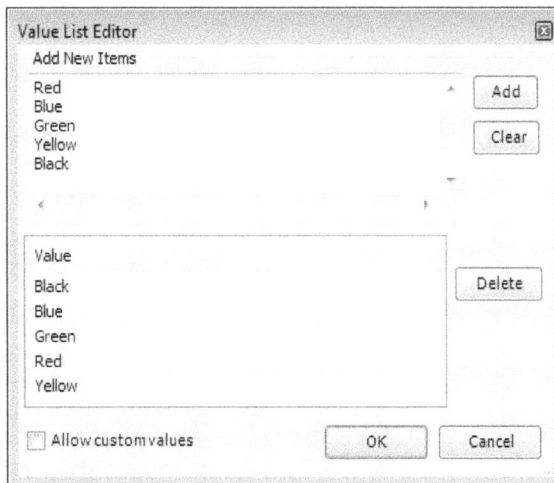

Figure 5–16

5. Close the Value List Editor and Parameters dialog boxes. Note that the screwdriver color does not update.

Task 3 - Add a rule to the assembly to control the screwdriver color.

A rule already exists in the **screwdriver_body** component that controls the color of the component. In the top-level assembly, you now need to set the **ScrewdriverColor** assembly user parameter so that it adjusts the color of the body at the part level.

1. In the *Manage* tab>iLogic panel, click 📜 (Add Rule) to add a rule.

2. Enter **Screwdriver_Color** as the name of the rule. The Edit Rule dialog box opens.

3. In the *Model* tab at the top of the dialog box, expand the **screwdriver_body** component. Select **User Parameters** in the browser list and double-click on the **Body_Color** parameter to insert it into the lower area of the editor.

4. Enter **=** at the end of the inserted information.

5. Select **User Parameters** for the top-level assembly and double-click on the **ScrewdriverColor** parameter to insert it into the lower area of the editor, as shown in Figure 5–17.

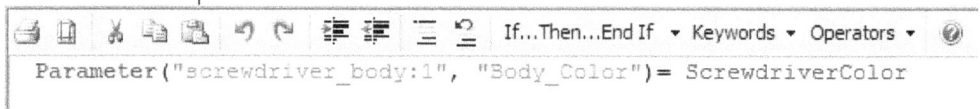

```
Parameter("screwdriver_body:1", "Body_Color")= ScrewdriverColor
```

Figure 5–17

6. Click **OK** to complete the rule. Note that the part color updates to the parameter value that is set in the assembly (i.e., Black).

7. Open the Parameters dialog box and modify the **ScrewdriverColor** parameter value to **Green** to verify that the body color updates. Close the Parameters dialog box.

The cap component does not change color because you need to incorporate it into the rule. The **screwdriver_cap** component is an iPart. Currently, as the instance is changed, the Model browser display name updates to show the version that is being used. A static name (occurrence name) is required for the cap component so that it consistently displays the same name regardless of the instance and this name is used in iLogic rules.

8. Locate the **screwdriver_cap** component in the Model browser. Click it once, then click it again to rename it. Enter **screwdriver_cap** as the new name, as shown in Figure 5–18.

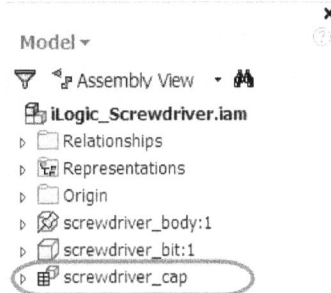

Figure 5–18

9. Right-click on **screwdriver_cap** in the Model browser and select **iProperties**. Select the *Occurrence* tab and note that the **screwdriver_cap** displays as the new occurrence name. This name is now static.

10. Change the **screwdriver_cap** iPart to the **Long Cap** variation. Note that the new name remains displayed in the Model browser regardless of the instance.

11. Return the **Standard Cap** instance to the display before continuing.

12. In the *Rules* tab of the iLogic browser, double-click on **Screwdriver_Color** to open the Rules editor.

13. In the editor portion of the dialog box, ensure that the cursor is on a new line, under the initial statement.

14. In the *Snippets* area, expand Components and double-click on **Color** to insert the snippet into the rule.

15. Enter **=** at the end of the inserted information.

16. Select **User Parameters** for the top-level assembly in the *Model* tab and double-click on the **ScrewdriverColor** parameter to insert it into the lower area of the editor, as shown in Figure 5–19.

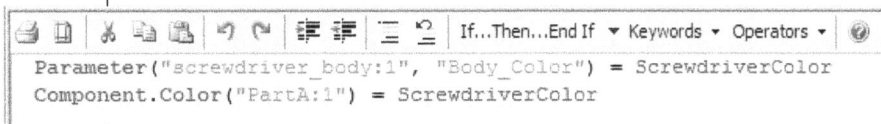

Figure 5–19

17. Highlight **PartA:1** in the editor. Select **screwdriver_cap** in the Model list, select the *Names* tab to the right of the list, and double-click on **screwdriver_cap** to replace the highlighted parameter in the rule. The rule displays as shown in Figure 5–20.

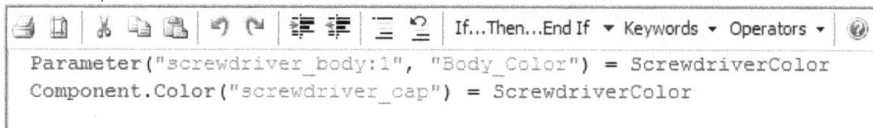

```
Parameter("screwdriver_body:1", "Body_Color") = ScrewdriverColor
Component.Color("screwdriver_cap") = ScrewdriverColor
```

Figure 5–20

18. Click **OK** to complete the rule. Using this method of assigning part color directly at the assembly level and passing it down to the part is more efficient than creating a rule in an iPart file. If created at the part level, color statements would have to be written for each instance of the iPart.

19. The cap component color should update. If not, update the assembly or verify that the rule was written correctly.

20. Open the Parameters dialog box and modify the **ScrewdriverColor** parameter value to **Black** to verify that the body and cap colors update.

21. Save the assembly and all of its files.

Task 4 - Add comments to the rule to describe the statements.

As the primary developer of a rule, you understand why all the statements are added. However, if anyone else reviews or has to update the model, additional guidance on the purpose of the statements can be helpful. In this task, you will add comment statements to the rule.

1. In the *Rules* tab of the iLogic browser, double-click on **Screwdriver_Color** to open the Edit Rule dialog box.

2. In the *Rule Editor* area of the dialog box, add an empty line above each of the two statements in the rule.

3. On line one, type **'Sets the Body_Color user parameter in the screwdriver body component to the value of the assembly's ScrewdriverColor parameter.**, as shown in Figure 5–21.

4. On line three, type **'Sets the color of the screwdriver cap component based on the ScrewdriverColor parameter value.**, as shown in Figure 5–21.

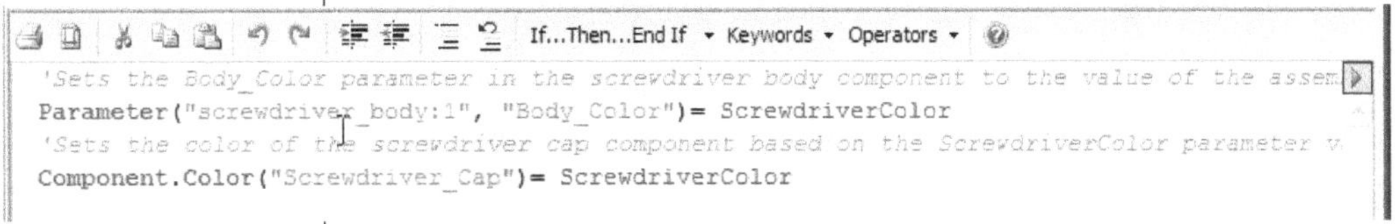

```
'Sets the Body_Color parameter in the screwdriver body component to the value of the assem
Parameter("screwdriver_body:1", "Body_Color")= ScrewdriverColor
'Sets the color of the screwdriver cap component based on the ScrewdriverColor parameter v
Component.Color("Screwdriver_Cap")= ScrewdriverColor
```

Figure 5–21

5. Click **OK** to complete the rule.

Task 5 - Add a rule to adjust the screwdriver size and type.

In this task you will suppress the **Screwdriver_Cap** component. To achieve this, you need to first create a Level of Detail Representation (LOD). Using a LOD, components can be suppressed in a rule. You will also rename constraints (relationships) to make them easier to identify when coding the suppression rules.

1. In the Model browser, expand the **Representations** node, right-click on **Level of Detail**, and select **New Level of Detail**.

2. Expand the **Level of Detail** node and rename the new LOD to **iLogic**.

3. In the Model browser, expand the **screwdriver_body** component. Rename the *Insert:1* constraint as **CapBodyInsert**.

4. Expand the **screwdriver_bit** component and rename its three constraints, as follows:

   - *Mate:1* to **BitCL**
   - *Angle:!* to **BitAngular**
   - *Mate:2* to **BitMate**

The Model browser displays as shown in Figure 5–22.

Figure 5–22

5. Add a new rule called **ScrewdriverSizeType**.

6. Click **If...Then...End If** to form the basis of the conditional rule.

7. Replace **My_Expression** with the **ScrewdriverType** user parameter.

8. Complete the statement by adding **= "Standard Multi-Bit"**, as shown in Figure 5–23.

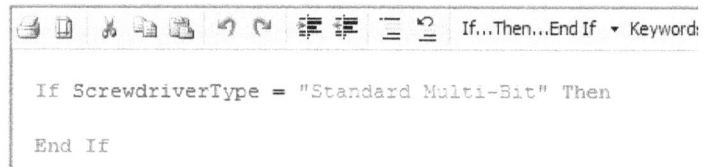

```
If ScrewdriverType = "Standard Multi-Bit" Then

End If
```

Figure 5–23

9. Insert the **EndType** and **Type** parameters from the **screwdriver_body** component. Set both of these parameters to **Multi-Bit** and **Standard**, respectively, as shown in Figure 5–24.

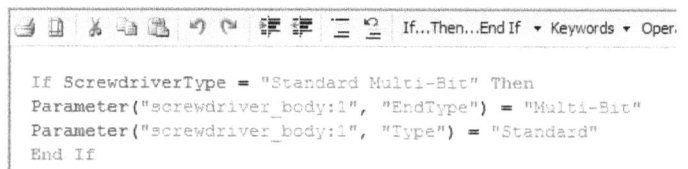

```
If ScrewdriverType = "Standard Multi-Bit" Then
Parameter("screwdriver_body:1", "EndType") = "Multi-Bit"
Parameter("screwdriver_body:1", "Type") = "Standard"
End If
```

Figure 5–24

10. Add a new line before the End If statement.

11. For the Standard Multi-Bit version of the assembly, the **Screwdriver_Cap** must be available (not suppressed). In the *Model* tab, right-click on the **screwdriver_Cap** component and select **Capture Current State** to set the component inclusion to **True**.

12. Add a new line before the End If statement.

13. In the *Model* tab, expand the **Screwdriver_Cap** component, right-click on the **CapBodyInsert** constraint, and select **Capture Current State** to include the constraint in the rule. The rule displays as shown in Figure 5–25.

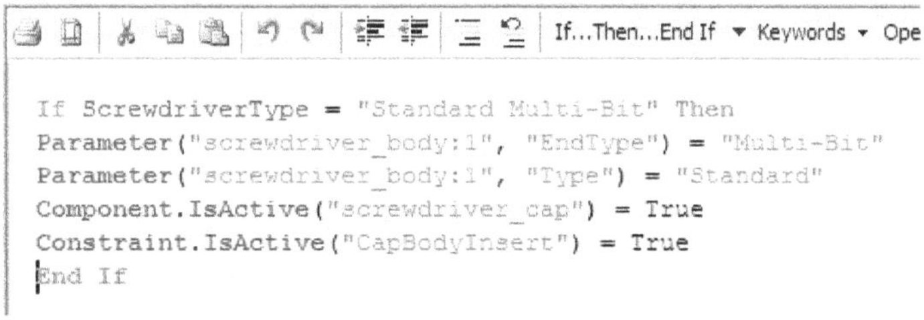

```
If ScrewdriverType = "Standard Multi-Bit" Then
Parameter("screwdriver_body:1", "EndType") = "Multi-Bit"
Parameter("screwdriver_body:1", "Type") = "Standard"
Component.IsActive("screwdriver_cap") = True
Constraint.IsActive("CapBodyInsert") = True
End If
```

Figure 5–25

14. Add a new line before the End If statement.

15. Expand **If...Then...End If** and select **ElseIf...Then**.

16. Copy and paste the four lines of code in the initial If statement and edit the code, as shown in Figure 5–26.

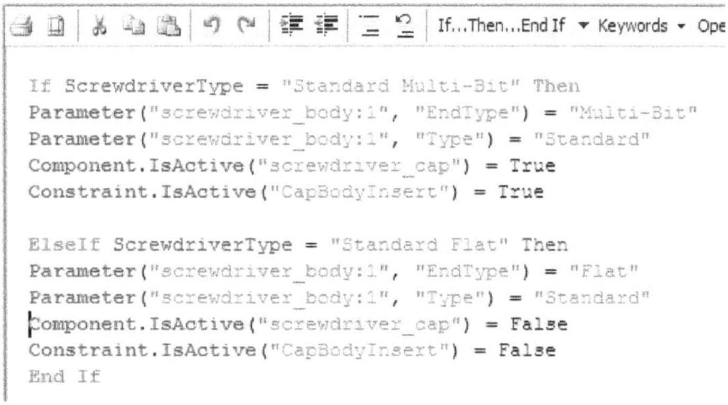

```
If ScrewdriverType = "Standard Multi-Bit" Then
Parameter("screwdriver_body:1", "EndType") = "Multi-Bit"
Parameter("screwdriver_body:1", "Type") = "Standard"
Component.IsActive("screwdriver_cap") = True
Constraint.IsActive("CapBodyInsert") = True

ElseIf ScrewdriverType = "Standard Flat" Then
Parameter("screwdriver_body:1", "EndType") = "Flat"
Parameter("screwdriver_body:1", "Type") = "Standard"
Component.IsActive("screwdriver_cap") = False
Constraint.IsActive("CapBodyInsert") = False
End If
```

Figure 5–26

17. As previously discussed, it is recommended that comments be added to the code to help describe the purpose of the statements. Add the comments shown in Figure 5–27.

```
If ScrewdriverType = "Standard Multi-Bit" Then
    'sets the user parameters of the screwdriver_body component
    Parameter("screwdriver_body:1", "EndType") = "Multi-Bit"
    Parameter("screwdriver_body:1", "Type") = "Standard"
    'ensures the cap and its constraint are not suppressed
    Component.IsActive("screwdriver_cap") = True
    Constraint.IsActive("CapBodyInsert") = True

ElseIf ScrewdriverType = "Standard Flat" Then
    'sets the user parameters of the screwdriver_body component
    Parameter("screwdriver_body:1", "EndType") = "Flat"
    Parameter("screwdriver_body:1", "Type") = "Standard"
    'ensures the cap and its constraint are suppressed (body is solid)
    Component.IsActive("screwdriver_cap") = False
    Constraint.IsActive("CapBodyInsert") = False
End If
```

Figure 5–27

To control the iPart version of the screwdriver_cap component, you need to use the **ChangeRow** function. Since **screwdriver_cap** is suppressed in the Flat version of the screwdriver, you only need to add code to the first If statement in the rule.

18. Add a new line before the ElseIf statement.

19. Expand the **iParts** node in the *Snippets* area. Double-click on **ChangeRow** to insert the snippet into the rule. Modify the statement as shown in Figure 5–28.

```
If ScrewdriverType = "Standard Multi-Bit" Then
    'sets the user parameters of the screwdriver_body component
    Parameter("screwdriver_body:1", "EndType") = "Multi-Bit"
    Parameter("screwdriver_body:1", "Type") = "Standard"
    'ensures the cap and its constraint are not suppressed
    Component.IsActive("screwdriver_cap") = True
    Constraint.IsActive("CapBodyInsert") = True
    'sets the version of the cap iPart
    iPart.ChangeRow("screwdriver_cap", "screwdriver_cap-01")

ElseIf ScrewdriverType = "Standard Flat" Then
    'sets the user parameters of the screwdriver body component
```

Figure 5–28

Code is now required to control the suppression of the **screwdriver_bit** component.

20. Capture the current state of the **screwdriver_bit** component and the three constraints used to constrain it, as shown in Figure 5–29.

```
If...Then...End If ▾ Keywords ▾ Operat

If ScrewdriverType = "Standard Multi-Bit" Then
'sets the user parameters of the screwdriver_body component
Parameter("screwdriver_body:1", "EndType") = "Multi-Bit"
Parameter("screwdriver_body:1", "Type") = "Standard"
'ensures the cap and its constraint are not suppressed
Component.IsActive("screwdriver_cap") = True
Constraint.IsActive("CapBodyInsert") = True
'sets the version of the cap iPart
iPart.ChangeRow("screwdriver_cap", "screwdriver_cap-01")
'ensure the bit and its constraint are not suppressed
Component.IsActive("screwdriver_bit:1") = True
Constraint.IsActive("BitCL") = True
Constraint.IsActive("BitAngular") = True
Constraint.IsActive("BitMate") = True

ElseIf ScrewdriverType = "Standard Flat" Then
'sets the user parameters of the screwdriver_body component
```

Figure 5–29

21. Copy the four lines of code that control the **screwdriver_bit** suppression into the ElseIf statement, and modify it as shown in Figure 5–30.

```
If...Then...End If ▾ Keywords ▾ Operators ▾
Component.IsActive("screwdriver_bit:1") = True
Constraint.IsActive("BitCL") = True
Constraint.IsActive("BitAngular") = True
Constraint.IsActive("BitMate") = True

ElseIf ScrewdriverType = "Standard Flat" Then
'sets the user parameters of the screwdriver_body component
Parameter("screwdriver_body:1", "EndType") = "Flat"
Parameter("screwdriver_body:1", "Type") = "Standard"
'ensures the cap and its constraint are suppressed (body is solid)
Component.IsActive("screwdriver_cap") = False
Constraint.IsActive("CapBodyInsert") = False
'ensure the bit and its constraint are not suppressed
Component.IsActive("screwdriver_bit:1") = False
Constraint.IsActive("BitCL") = False
Constraint.IsActive("BitAngular") = False
Constraint.IsActive("BitMate") = False
```

Figure 5–30

22. Click **OK** to complete the rule. The model updates to reflect a Standard flat black version of the screwdriver.

Task 6 - Verify the ScrewdriverSizeType rule.

In this task you verify the **ScrewdriverSizeType** rule to ensure that it captures the model's design intent.

1. Open the Parameters dialog box.

2. In the *ScrewdriverType* drop-down list, change the value to **Standard Multi-Bit**. The cap component and bit are now visible.

3. Change the *ScrewdriverType* to **Stubby Flat**.

4. The model geometry does not change because nothing is written in the rule about this model variation. Edit the **ScrewdriverSizeType** rule to add the code for the other two versions. Edit the code, as shown in Figure 5–31.

```
Constraint.IsActive("BitMate") = False

ElseIf ScrewdriverType = "Stubby Flat" Then
'sets the user parameters of the screwdriver_body component
Parameter("screwdriver_body:1", "EndType") = "Flat"
Parameter("screwdriver_body:1", "Type") = "Stubby"
'ensures the cap and its constraint are suppressed (body is solid)
Component.IsActive("screwdriver_cap") = False
Constraint.IsActive("CapBodyInsert") = False
'ensure the bit and it constraints are not suppressed
Component.IsActive("screwdriver_bit:1") = False
Constraint.IsActive("BitCL") = False
Constraint.IsActive("BitAngular") = False
Constraint.IsActive("BitMate") = False

ElseIf ScrewdriverType = "Stubby Multi-Bit" Then
'sets the user parameters of the screwdriver_body component
Parameter("screwdriver_body:1", "EndType") = "Multi-Bit"
Parameter("screwdriver_body:1", "Type") = "Stubby"
'ensures the cap and its constraint are not suppressed
Component.IsActive("screwdriver_cap") = True
Constraint.IsActive("CapBodyInsert") = True
'sets the version of the cap iPart
iPart.ChangeRow("screwdriver_cap", "screwdriver_cap-03")
'ensure the bit and it constraints are not suppressed
Component.IsActive("screwdriver_bit:1") = True
Constraint.IsActive("BitCL") = True
Constraint.IsActive("BitAngular") = True
Constraint.IsActive("BitMate") = True

End If
```

Figure 5–31

5. Click **OK** to complete the rule.

6. Update the model so that the geometry reflects that the model has been previously set to **Stubby Flat**.

7. Change the **ScrewdriverType** to **Stubby Multi-Bit** and **ScrewdriverColor** to **Red**. Close the Parameters dialog box.

8. Save the assembly and close all of the files.

Task 7 - (Optional) Add a rule to control the bit type used in the multi-bit screwdriver versions.

If time permits, add a new multi-list text parameter to the assembly that controls the type of bit that can be used. Once the parameter is created, add a rule to control the bit type based on the value of the new parameter.

Chapter Review Questions

1. An iLogic rule must be included at the part level in order to control the suppression state of the entire component in an assembly.

   a. True

   b. False

2. Which of the following tabs are **only** available in the Edit Rule dialog box when creating a rule in an assembly file? (Select all that apply.)

   a. Model

   b. File Tree

   c. Files

   d. Options

   e. Search and Replace

   f. Wizards

3. Which of the following lines of code assigns a parameter from a top-level assembly to populate a parameter that exists at the part level of an assembly component?

   a. Parameter("Part:1", "param_name_part") = Parameter

   b. Parameter = Parameter("Part:1", "param_name_part")

4. Which of the following is required if an assembly rule is to control component suppression?

   a. View Representation

   b. Positional Representation

   c. Level of Detail Representation

5. When a component is suppressed using an **IsActive** component function, its BOM Structure is set to which of the following options?

   a. **Normal**

   b. **Inseparable**

   c. **Purchased**

   d. **Phantom**

   e. **Reference**

6. A component called **Fastener** is not suppressed and is visible in a top-level assembly model. Based on the following rule, which option best describes the model?

```
If Component.Visible("Fastener") = True Then
Feature.IsActive("Center_Hole") = True
Hole_dia = 1
Else If Component.Visible("Fastener") = False Then
Feature.IsActive("Center_Hole") = False
End If
```

   a. Hole_dia = 1 and the Center_Hole feature is suppressed

   b. Hole_dia = 1 and the Center_Hole feature is displayed

   c. Hole_dia is unknown and Center_Hole feature is suppressed.

   d. Hole_dia is unknown and Center_Hole feature is displayed.

7. iProperty calculations update to reflect that an assembly component's visibility status has been set such that it is not visible in the assembly.

   a. True

   b. False

8. Which of the following lines of code displays the correct syntax for a function that replaces a component with another instance from its iPart table? The iPart factory member is fully defined with no custom parameter values required.

   a. Component.Replace(Base", "Base_1.ipt", True)

   b. Component.Replace(Base", "Base.ipt", True, 1)

   c. Component.Replace(Base", "Base.ipt", True, 1, 20)

   d. Component.Replace("Piston:1", "Piston.iam<Low>", True)

9. Constraints cannot be renamed in the Model browser. They must maintain their original naming convention in order to be used in an assembly iLogic rule.

   a. True

   b. False

10. Which of the following best describes the difference between assembling a standard component and assembling an iLogic component? (Select all that apply.)

   a. Before assembling an iLogic component you must save the assembly in which you are assembling the iLogic component.

   b. When assembling an iLogic component, the component is immediately opened in the active assembly and you can constrain it as required.

   c. The Place iLogic Component dialog box enables you to define values for parameters in the assembly that are used to drive the iLogic rules. You do not have to change parameters using the Parameters dialog box to configure the iLogic component for use in the assembly.

   d. When an iLogic component is assembled, a copy of the source model is generated and stored in a new folder with the same name as the assembly it is being assembled into. This folder is located at the same level as the active assembly.

Command Summary

Button	Command	Location
	Add Rule	• **Ribbon:** *Manage* tab>iLogic panel • **iLogic browser:** Right-click
	Place iLogic Component	• **Ribbon:** *Assemble* tab>expanded Component panel

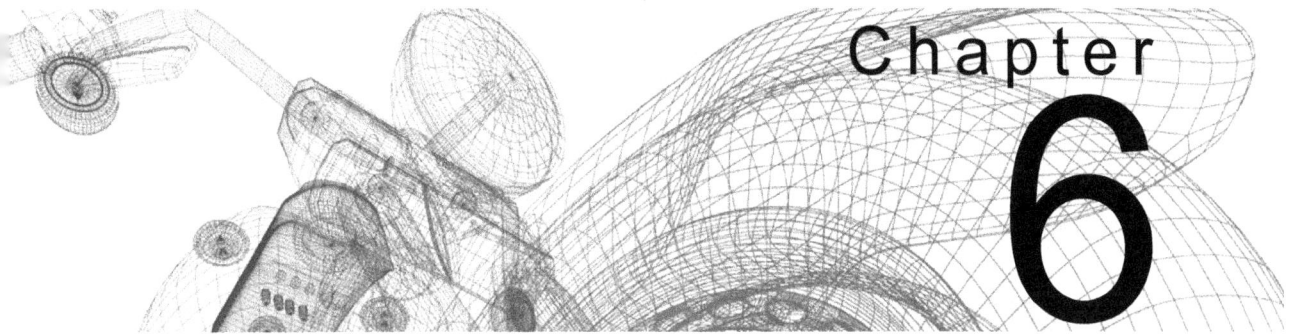

Drawing Rules and Functions

Using iLogic rules in a drawing document, you can automate many actions that might have to be completed manually to cleanup a drawing after a model change. Some actions that can be used in iLogic drawing rules affect sheet sizes, title blocks, borders, view positioning, scaling, and suppression.

Learning Objectives in this Chapter

- Describe which elements in an Autodesk® Inventor® drawing document can be controlled using drawing specific functions.
- Assign the active drawing sheet and read its name, size, width, and height using iLogic functions.
- Change the size of the active sheet and determine whether tables remain positioned along the top and right edges using an iLogic function.
- Use an iLogic function to assign another drawing that can be used as a resource for title blocks and borders.
- Use an iLogic function to change the title block and border assigned in the current drawing sheet.
- Read the width and height of an existing view on a drawing sheet using an iLogic function.
- Change the scale of an existing view on a drawing sheet using an iLogic function.
- Change the position of an existing view on a drawing sheet relative to sheet edges or to other views using an iLogic function.
- Reattach balloons in a drawing view using an iLogic function.
- Suppress a view on a drawing sheet using an iLogic function.
- Read the drawing model's name that is used in a drawing document or view using an iLogic function.

iLogic Workflow

Figure 6–1 shows the overall suggested workflow for the iLogic tools in the Autodesk Inventor software. The horizontal line at the top represents the high-level workflow and each of their sub-steps are detailed vertically below them. The highlighted column represents the content discussed in the current chapter.

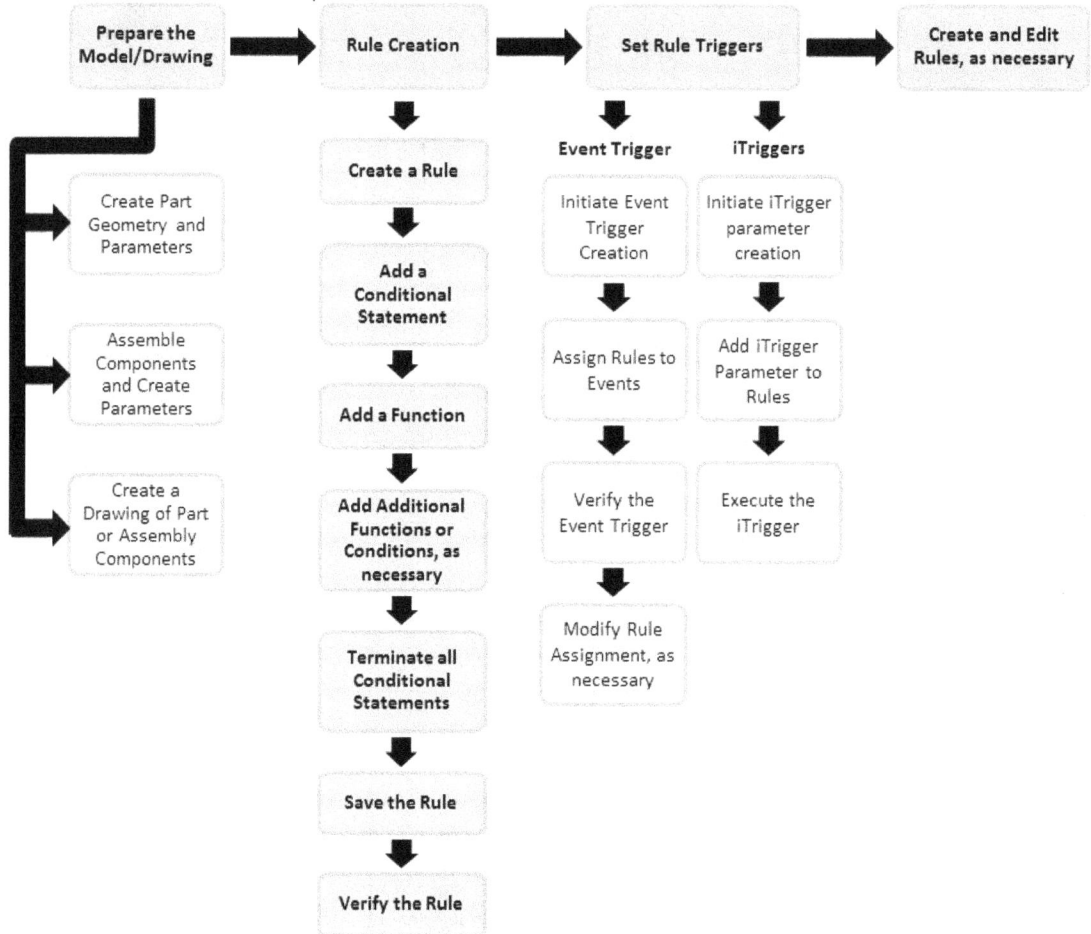

Prepare the Model/Drawing	Rule Creation	Set Rule Triggers	Create and Edit Rules, as necessary

Prepare the Model/Drawing
- Create Part Geometry and Parameters
- Assemble Components and Create Parameters
- Create a Drawing of Part or Assembly Components

Rule Creation
- Create a Rule
- Add a Conditional Statement
- Add a Function
- Add Additional Functions or Conditions, as necessary
- Terminate all Conditional Statements
- Save the Rule
- Verify the Rule

Event Trigger
- Initiate Event Trigger Creation
- Assign Rules to Events
- Verify the Event Trigger
- Modify Rule Assignment, as necessary

iTriggers
- Initiate iTrigger parameter creation
- Add iTrigger Parameter to Rules
- Execute the iTrigger

Figure 6–1

6.1 Drawing Rules

The parametric behavior of the Autodesk Inventor software ensures that changes to the model are immediately reflected in the drawing. However, some documentation specific changes (sheet sizes, views, title blocks, etc.) are generally required due to a model change but are not automatically done. Using iLogic rules in a drawing document, you can automate many of these actions. Some actions that can be used in the iLogic rules in a drawing document include:

Additionally, advanced functions can be used so the document is used as a template. Rules can be assigned to select a sheet size or a title block from scratch and assign a model.

- Controlling an active sheet's size.

- Controlling an active sheet's view placement, size, or scale.

- Controlling an active sheet's title block or border.

- Reattaching balloons to a drawing view.

- Controlling view suppression in an active drawing sheet.

- Controlling the visibility of layers in an active drawing sheet.

In the example shown in Figure 6–2, iLogic was used to control the title block that is displayed, the sheet size, and the view scale, and placement based on changes to the length of the drawing model.

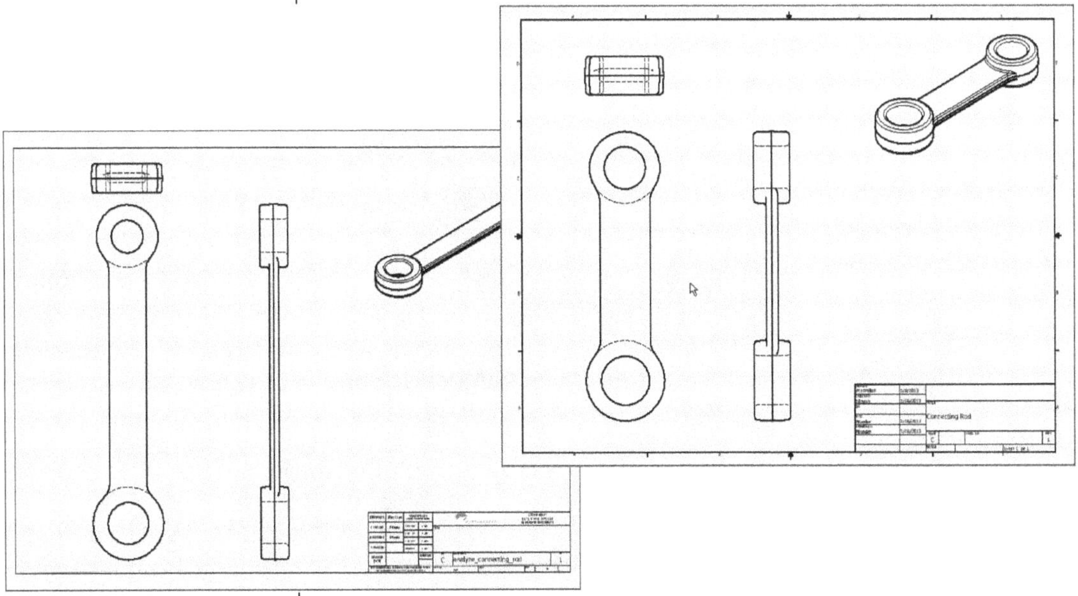

Figure 6–2

The workflow for implementing iLogic rules in a drawing document can be broken down into four steps. Use the following general steps to create iLogic rules in a drawing:

1. Prepare the drawing.
2. Rule Creation.
3. Set rule triggers, as necessary.
4. Create and edit rules, as necessary.

Figure 6–3 highlights the steps graphically. Additional in-depth information is included on each of these steps as you progress through this student guide.

| Prepare the Drawing | ➡ | Rule Creation | ➡ | Set Rule Triggers | ➡ | Create and Edit Rules, as necessary |

Figure 6–3

Step 1 - Prepare the drawing.

If the drawing is intended as a blank drawing template, views and title block can be left off the drawing and can be added using the rules.

To prepare a drawing so that it can be automated to modify drawing sheets, views, or the title block, begin by ensuring that all the drawing components exist in the drawing. Also verify that all possible views and sheets that might be required exist and are positioned as required. Additionally, if you want to switch the title blocks with iLogic rules, ensure that a drawing title block is assigned and any other title block(s) that might be required are located in the drawing resources of the current drawing.

Step 2 - Rule Creation.

In the *Manage* tab>iLogic panel, click (Add Rule) to begin the creation of a new rule in the drawing document. The Edit Rule dialog box opens and is similar to that used for part rules. The steps for creating drawing rules are the same as those for creating rules in part models. However, the functions vary depending on the required drawing action.

Once rules are created, verify that they are working as expected by flexing the drawing model or drawing parameters, as required. Rules should also be explicitly run to verify that they capture the design intent.

Step 3 - Set rule triggers, as necessary.

Rule triggers enable you to define when a rule is run. iLogic provides a list of event triggers to which the established rules are assigned. The list of triggers vary slightly depending on whether a part, assembly, or drawing is active. Each trigger provides you with standard functions commonly used, such as, before a document save, when a document is closed, or when part geometry is changed. An iTrigger can also be used to trigger rules by adding a user parameter to the document that in turn launches any rules that contain it.

Step 4 - Create and edit rules, as necessary.

Continue to add or edit rules, as required. A complete list of the rules can be reviewed in the iLogic browser. The order rules are listed in the iLogic browser can affect the drawing. Drag and drop rules in the iLogic browser to verify that their order captures the design intent.

If changes are required in a rule, right-click the rule name in the iLogic browser and select **Edit Rule**. Alternatively, you can double-click the rule to open the Edit Rule dialog box.

6.2 Sheet Functions

The *Drawing* category in the *System* tab in the *Snippets* area provides functions for creating rules that control sheets, views, and title blocks. The expanded list of Drawing functions specific to Sheet functions is shown in Figure 6–4.

Hover the cursor over a function to display a short help line indicating its syntax and how it can be used in a rule.

*The **ThisDrawing** and **ActiveSheet** functions are reference functions that can be used in lines of code to call out the current drawing and active sheet in that drawing. These functions exist in many of the other functions in the list.*

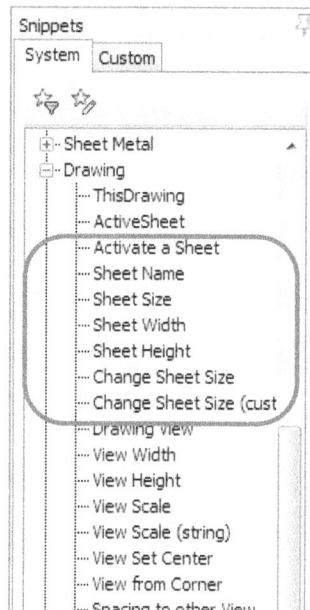

Figure 6–4

Sheet functions enable you to create rules that reads the sheet name and size, as well as change the sheet size of the current drawing.

Activate a Sheet

The **Activate a Sheet** function assigns the drawing's active sheet to that assigned in the rule. It can be incorporated in the code to set a specific drawing sheet as the active sheet to ensure that the functions that follow it are applied to it. The syntax for the **Activate a Sheet** function code is as follows. The Assembly Isometric:1 sheet is active once this code is run.

```
ActiveSheet = ThisDrawing.Sheet("Assembly Isometric:1")Sheet Name and Sizes
```

Sheet Information

The **Sheet Name**, **Sheet Size**, **Sheet Width**, and **Sheet Height** functions enable you to read the name and sizes of the active sheet. The code's syntax for each of these function is as follows:

```
ActiveSheet.Name
ActiveSheet.Size
ActiveSheet.Width
ActiveSheet.Height
```

Each of these could be used as a condition in a rule's conditional statement.

Change Sheet Size

The **Change Sheet Size** functions enable you to assign a new sheet size to the active sheet. The function includes the new sheet size as the first parameter in the brackets and **MoveBorderItems** as an optional parameter. The code's syntax can be either of the following:

A colon is required in the operator for the MoveBorderItems value.

```
ActiveSheet.ChangeSize("A3")
ActiveSheet.ChangeSize("A3", MoveBorderItems := True)
```

• If **MoveBorderItems** is set to **False**, tables/parts lists located on the top or right edge of the sheet are not moved to the edge of the new sheet. If set to True, tables/parts lists located on these edges are moved to the edge of the new sheet. This parameter does not control placement of tables along the left or bottom.

The **Change Sheet Size (custom)** function is a slight variation of **Change Sheet Size**, where you provide custom dimension sizes for a custom sheet. The units of the sheet are the same as the document units. The code's syntax for the **Change Sheet Size (custom)** function can be either of the following:

A colon is required in the operator for the MoveBorderItems value.

```
ActiveSheet.ChangeSize(7, 5)
ActiveSheet.ChangeSize(7, 5, MoveBorderItems := True)
```

6.3 Title Block Functions

The *Drawing* category in the *System* tab in the *Snippets* area provides access to functions that can be used in creating rules that control sheets, views, and title blocks. The expanded list of Drawing functions specific to Title Block functions is shown in Figure 6–5.

Hover the cursor over a function to display a short help line indicating its syntax and how it can be used in a rule.

*The **ThisDrawing** and **ActiveSheet** functions are reference functions that can be used in lines of code to call out the current drawing and active sheet in that drawing. These functions exist in many of the other functions in the list.*

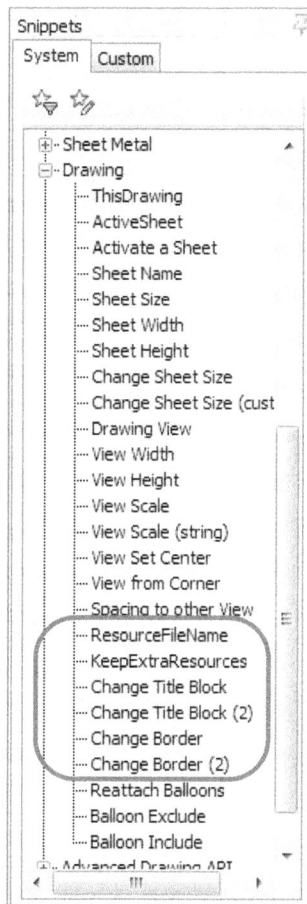

Figure 6–5

Drawing Resources

Title blocks and borders used in a drawing must exist in the Drawing Resources for the drawing, as shown in Figure 6–6. Multiple Drawing Resources can exist even if they are not all used.

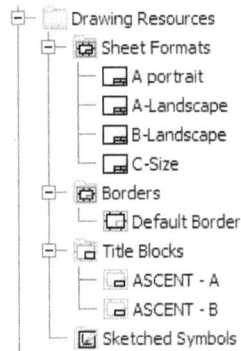

Figure 6–6

When controlling a title block or border change in an iLogic drawing rule, any new title block or border should also exist in the associated Drawing Resource for the drawing. However, you can also assign an alternate drawing file that can be used as a resource. Once assigned, the resource drawing is referenced in situations where a replacement title block or border is not found in the active drawing.

- Use the **ResourceFileName** function to define the resource file in a drawing rule. The syntax for defining a resource file is as follows, where <DrawingResources1.idw> is the name of the resource drawing.

```
ThisDrawing.ResourceFileName = "<DrawingResources1.idw>"
```

- Use the **KeepExtraResources** function to control whether resources used from the resource file are copied into the current drawing or not. If the **KeepExtraResources** function is set to *True*, it is copied; if set to *False*, it is copied until the resource itself is replaced and then it is deleted.

```
ThisDrawing.ResourceFileName = "<DrawingResources1.idw>"
ThisDrawing.KeepExtraResources = True
```

Change Title Block

The **Change Title Block** function enables you to assign a new title block to the active sheet. The syntax for the code is as follows, where <New Title Block Name> represents the name of the new title block being assigned to the sheet. The new title block must be located in the drawing resources of the current drawing or in a referenced resource drawing.

```
ActiveSheet.TitleBlock = "<New Title Block Name>"
```

The **Change Title Block (2)** function enables you to expand the Change Title Block function to incorporate prompted entry for the title block information. The syntax for this code is similar to the original function with the addition of the prompted entry fields.

```
ActiveSheet.SetTitleBlock("<New Title Block Name>",
"promptedEntry1", "promptedEntry2", "promptedEntry3")
```

- The prompted entry fields correspond to the text that is added to the title block as Prompted Entry that is listed in the Format Text dialog box.

Change Border

The **Change Border** function enables you to assign a new border to the active sheet. The syntax for the code is as follows, where <New Border Name> represents the name of the new border being assigned to the sheet. The new border must be located in the drawing resources of the current drawing or in a referenced resource drawing.

```
ActiveSheet.Border = "<New Border Name>"
```

The **Change Border (2)** function enables you to expand on the Change Border function to incorporate prompted entry for border information. The syntax for this code is similar to the original function with the addition of the prompted entry fields.

```
ActiveSheet.Border("<New Border Name>", "promptedEntry1",
"promptedEntry2", "promptedEntry3")
```

- The prompted entry fields correspond to text that is added to the border as Prompted Entry that is listed in the Format Text dialog box.

6.4 View Functions

The *Drawing* category in the *System* tab in the *Snippets* area provides access to functions that can be used in creating rules that control sheets, views, and title blocks. The expanded list of Drawing functions specific to View functions is shown in Figure 6–7.

Hover the cursor over a function to display a short help line indicating its syntax and how it can be used in a rule.

*The **ThisDrawing** and **ActiveSheet** functions are reference functions that can be used in lines of code to call out the current drawing and active sheet in that drawing. These functions exist in many of the other functions in the list.*

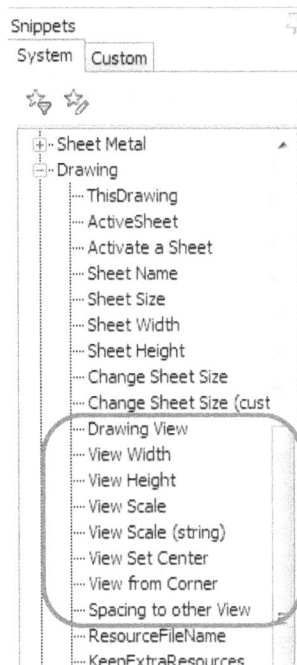

Figure 6–7

In general, the View functions enable you to create rules that change a view's size and position on the active sheet. In addition, there are a few functions that can be used to cleanup balloons that are associated with views in a drawing.

Drawing View

The **Drawing View** function can be used in a drawing rule to assign the active view on the active sheet. This function is generally used in combination with other coding elements and can be incorporated in the code to set a specific drawing view as the active view to ensure that the functions that follow are applied to it. The syntax for the **Drawing View** function code is as follows. The ViewNumber element identifies the view name that is listed in the Model browser or by a view label if a label has been displayed for the view.

```
ActiveSheet.View("ViewNumber")
```

View Size

The **View Width** and **View Height**, Sheet functions enable you to read the width and height of the active view. The syntax for each of these functions is as follows. The initial portion of the code identifies the drawing view that is being read and the last element identifies whether the width or height dimension is required.

```
ActiveSheet.View("VIEW1").Width
ActiveSheet.View("VIEW1").Height
```

Reading the width and height of a view can be incorporated into a rule to set a condition on a scale change that might be required if the model becomes too large for the sheet.

View Scale

The two **View Scale** functions enable you to set the scale value for an active view. The syntax for each of these functions is as follows.

```
ActiveSheet.View("VIEW1").Scale
ActiveSheet.View("VIEW1").ScaleString
```

The initial portion of the code identifies the drawing view that is being controlled and the last element identifies whether the scale value for the change is entered as a number or as a text string, as shown below.

```
ActiveSheet.View("VIEW1").Scale = 2
ActiveSheet.View("VIEW1").ScaleString = 2:1
```

View Positioning

The **View Set Center**, **View from Corner**, and **Spacing to other View** functions can be used to customize where views are placed in a sheet. These functions are important when changes are made to a model that affect the view size. Consider using these functions to clean up the drawing so that the views do not overlap.

- The **View Set Center** function enables you to reposition the center of the active view at specific X and Y coordinates on the sheet. The units should be consistent with the units of the drawing document. The syntax for this function is as follows.

```
ActiveSheet.View("VIEW1").SetCenter(centerX, centerY)
```

The initial portion of the code identifies the drawing view that is being moved and the last element identifies the coordinates to which the center of the view is positioned. The coordinates are always measured from the lower left corner with this function.

*This function is a good alternative to the **View Set Center** function if the coordinates need to be measured from a corner other than the lower left.*

- The **View from Corner** function enables you to reposition the the active view. This function enables you to control which sheet corner the defined X and Y coordinates are measured from. In addition, instead of positioning the view relative to the center of the view, it positions it relative to the view corner that is closest to the sheet corner specified in the code. The units should be consistent with the units of the drawing document. The syntax for this function is as follows.

  ```
  ActiveSheet.View("VIEW1").SetSpacingToCorner(distanceX,
  distanceY, SheetCorner.TopLeft)
  ```

  The initial portion of the code identifies the drawing view that is being moved and the last element identifies the coordinates to which the view is being moved to and the corner from which the coordinates are measured. The following is the syntax used when defining the sheet corners.

  ```
  SheetCorner.BottomLeft
  SheetCorner.TopLeft
  SheetCorner.BottomRight
  SheetCorner.TopRight
  ```

- The **Spacing to other View** function enables you to reposition one view relative to another adjacent view. The value that is assigned to this function identifies the spacing between view edges in the horizontal or vertical direction. This function is used to ensure that the views remain positioned relative to one another and regardless of changes to the model size. The units should be consistent with the units of the drawing document. The syntax for this function is as follows.

  ```
  ActiveSheet.View("VIEW2").SpacingBetween("VIEW1") = 1.2
  ```

  The initial portion of the code identifies the active drawing view that is being moved and the last element identifies the view relative to which the active view is being moved. The value entered at the end of the function defines the distance being maintained between the two views. Enter a positive value to position the view to the right or above the reference view. A negative value moves the active view below or to the left of the reference view.

Drawing Balloons

*The **Reattach Balloons** function requires you to set the **Preserve Orphaned Annotations** option in the drawing's Document Settings. If it is not already set, then the function sets it.*

For assembly drawings, balloons might be included in a view to help identify its components. The **Reattach Balloons**, **Balloon Exclude**, and **Balloon Include** functions enable you to incorporate code in a drawing rule that deals with the balloons in a view.

- The **Reattach Balloons** function checks the active view, verifies if there are any unattached balloons in the view, and reattaches them, if possible. The first choice for attachment is a component at the arrowhead location. Following that, the closest component without a balloon is used. If the unattached balloon cannot be attached automatically, it is moved to a hidden layer. The syntax for this function is as follows.

    ```
    ActiveSheet.View("VIEW1").Balloons.Reattach
    ```

    The initial portion of the code identifies the drawing view that is being verified for unattached balloons. The last element identifies the balloons that are being reattached.

- The **Balloon Include** and **Balloon Exclude** functions are used in conjunction with the **Reattach Balloons** function. In general, they enable you to specify components in the view that can be used for balloon reattachment or specify those that should never have balloons attached to them. To correctly include or exclude components from reattaching, these functions must exist before the **Reattach Balloons** function is listed in the code list.

    Using the **Balloon Include** function, you can provide a list of components that should have balloons attached. Using the **Balloon Exclude** function, you can provide a list of components that should not have balloons attached to them.

    In the following example, Part1, Part2, and Part3 can have balloons attached, while Part4 cannot attach balloons to it. Note how the include and exclude statements are listed before the **Reattach Balloons** function.

    ```
    ActiveSheet.View("VIEW1").Balloons.AttachToComponent("Part1:1")
    ActiveSheet.View("VIEW1").Balloons.AttachToComponent("Part2:1")
    ActiveSheet.View("VIEW1").Balloons.AttachToComponent("Part3:1")
    ActiveSheet.View("VIEW1").Balloons.DoNotAttachToComponent("Part4:1")
    ActiveSheet.View("VIEW1").Balloons.Reattach
    ```

6.5 Advanced Functions

The *Advanced Drawing API* category in the *System* tab in the *Snippets* area provides additional drawing functions. The expanded list of Advanced Drawing API functions are shown in Figure 6–8.

Hover the cursor over a function to display a short help line indicating its syntax and how it can be used in a rule.

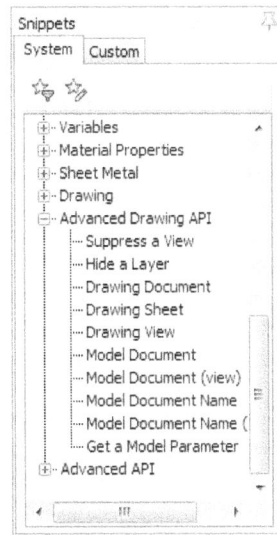

Figure 6–8

The Advanced Drawing API functions provide additional functions that can be used to suppress views and read drawing model information for use in drawing rules. Some of the functions in this category are advanced API functions and are not covered in this student guide.

View Suppression

The **Suppress a View** function is used to suppress views that are not required for certain drawing model configurations or might not be required for a specific drawing deliverable. The syntax for this function is as follows.

ActiveSheet.View("VIEW2").View.Suppressed = True

Define the view name that is to be suppressed in the brackets and set the function to True to suppress it. Set the function equal to False if the view is to be displayed.

Hide a Layer

The **Hide a Layer** function is used to hide defined layers in a drawing view. The syntax for this function is as follows.

```
ThisDrawing.Document.StylesManager.Layers("LayerName").
Visible = False
```

The initial portion of the function identifies that the layer exists in the Styles Manager of the current drawing document. In the bracketed area, enter the name of the layer as it is listed in the Style and Standard Editor. In the following example, the code is used to hide all drawing dimensions in the document. Setting the value to **True**, displays the layer.

```
ThisDrawing.Document.StylesManager.Layers("Dimension (ANSI)").
Visible = False
```

Model Documents

When creating a drawing rule, you might want to refer to the drawing model when creating the rule. For example, when creating a conditional statement, you might need to refer to a parameter in the drawing model. To do this, you must identify the model name. The following Model Document functions can be used in these situations.

- The **Model Document Name** function reads the value for the first model used in the drawing and assigns it to the variable called *modelName*. This function is used when calling out parameters in the drawing model.

  ```
  modelName = IO.Path.GetFileName(ThisDrawing.
  ModelDocument.FullFileName)
  ```

- If multiple drawing models exist in the drawing, use the **Model Document Name (view)** function to identify the view name for which you require the drawing model. In this following statement, the drawing model that was used to create VIEW3 is used to populate the modelName variable. This function is used when calling out parameters in the drawing model.

  ```
  modelName = IO.Path.GetFileName(ActiveSheet.View("VIEW3").
  ModelDocument.FullFileName)
  ```

Model Parameters

As its name implies, the **Get a Model Parameter** function can be used to get the value of a specific parameter from the drawing model (*modelName*) for use in a drawing rule.

```
dwgParam = Parameter(modelName, "Style")
```

Practice 6a

Automating Drawing Changes

Practice Objectives

- Create a drawing rule and specify the drawing sheet that is controlled by the rule.
- Create a drawing rule variable that represents the drawing model's name.
- Incorporate drawing functions into a rule to scale and reposition views on a drawing sheet.
- Incorporate drawing functions into a rule that defines a reference drawing as a source for Drawing Resources.
- Incorporate drawing functions into a rule that controls whether the drawing resources from a reference drawing are permanently or temporarily stored in the current drawing.
- Incorporate drawing functions into a rule that replaces title blocks and borders on a drawing sheet.

In this practice you work with an Autodesk Inventor drawing file to create an iLogic drawing rule that controls the placement of views on the drawing sheet and which title block and border are displayed.

Task 1 - Open and investigate the working models.

In this task, you will investigate the assembly and drawing file that you will be working with. In the assembly, a parameter has been created that enables you to select between three different size configurations. By changing the parameter in the assembly, the drawing that was created based on the smallest configuration, experiences overlapping views, and scaling issues when it is updated.

1. Open **Vise.idw** from the *Drawing* folder. The drawing contains four views, as shown in Figure 6–9.

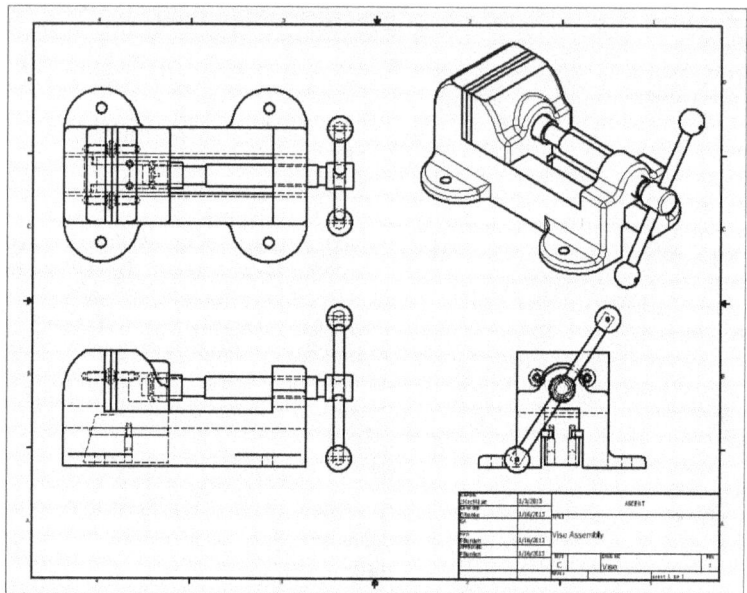

Figure 6–9

2. Open **Vise.iam**. This is the model that is used in the drawing file.

3. Open the **Vise_Size** rule. This rule drives the size of the **Base_vise** and **Vise_screw** components in the assembly. Review the rule. The **Vise_Size** user parameter was added to the model and its value determines the sizes in the two components.

4. In the assembly, open the Parameters dialog box and change the **Vise_Size** parameter value to **Medium**.

5. Activate the drawing file and note how views now overlap the border of the sheet. The original drawing was created with the Small configuration of the model

6. Return to the assembly file, open the Parameters dialog box and change the **Vise_Size** parameter value to **Large**.

7. Activate the drawing file. Note that the views still overlap the border of the sheet and the scale of the views is too big for the sheet.

8. Return to the assembly file and change the **Vise_Size** parameter value to **Small**.

Task 2 - Create a drawing rule to position and scale views based on specific conditions.

In this task, you will create a drawing rule to reposition and scale the drawing views if the **Vise_Size** parameter value is set to anything other than Small.

1. Return to the drawing file.

2. In the *Manage* tab>iLogic panel, click ▤ (Add Rule) to add a rule.

3. Enter **Drawing_Cleanup** as the name of the rule. The Edit Rule dialog box opens.

4. In the *Rule Editor* area of the dialog box, enter **Controls view scale and positioning for Sheet 1** as a statement that identifies the intention of the rule. With the cursor still on this line of code, click ▤ (Comment) to comment out the statement so that it is not read by the rule, but is meant for information.

5. The drawing is currently only a single sheet. The code that is to be added will only pertain to this sheet. Expand the *Drawing* category in the *Snippets* area. Double-click on **Activate a Sheet** to insert the snippet into the rule. Modify the statement so that the correct sheet name is identified in the brackets, as shown in Figure 6–10. Refer to the Model browser to obtain the correct name of the sheet (Sheet:1).

6. In a later task, a conditional statement is used that requires reading data in the drawing model. The drawing model name must be identified in the rule. Expand the *Advanced Drawing API* category in the *Snippets* area. Double-click on Model Document Name to insert the snippet into the rule, as shown in Figure 6–10. This sets the *modelName* variable to that of the drawing model for use later in the rule.

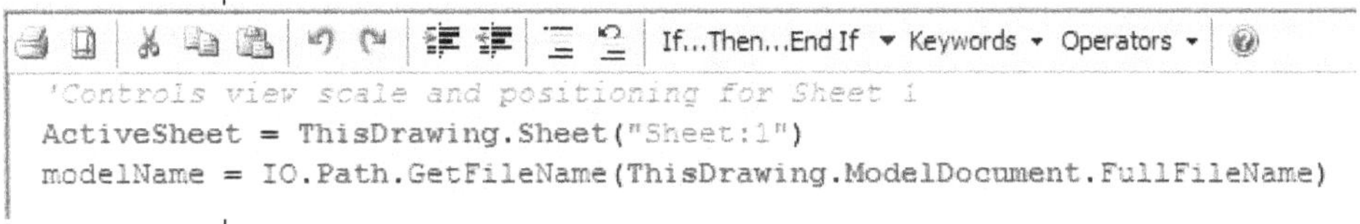

```
        If...Then...End If ▼  Keywords ▼  Operators ▼   ⑦
'Controls view scale and positioning for Sheet 1
ActiveSheet = ThisDrawing.Sheet("Sheet:1")
modelName = IO.Path.GetFileName(ThisDrawing.ModelDocument.FullFileName)
```

Figure 6–10

7. Click **If...Then...End If** to form the basis of the conditional rule.

8. Replace **My_Expression** with a statement that reads the **Vise_Size** parameter value in the drawing model (**Vise.iam**) and evaluates if it is equal to **Small**. Enter the expression, as shown in Figure 6–11.

9. Add a new line before the End If statement.

10. Expand **If...Then...End If** and select **ElseIf...Then**.

11. Replace **My_Expression** with a statement that also reads the **Vise_Size** parameter value in the drawing model (Vise.iam) and evaluates if it is equal to **Medium**. Enter the expression, as shown in Figure 6–11.

12. Copy and paste or recreate an additional **ElseIf...Then** statement for the **Large Vise_Size**. Enter the expression, as shown in Figure 6–11.

```
If...Then...End If  ▾  Keywords  ▾  Operators  ▾

'Controls view scale and positioning for Sheet 1
ActiveSheet = ThisDrawing.Sheet("Sheet:1")
modelName = IO.Path.GetFileName(ThisDrawing.ModelDocument.FullFileName)


If Parameter(modelname, "Vise_Size") = "Small" Then


ElseIf Parameter(modelname, "Vise_Size") = "Medium" Then


ElseIf Parameter(modelname, "Vise_Size") = "Large" Then


End If
```

Figure 6–11

13. Click **OK** to complete the rule. Ensure that the rule closes successfully. This verifies that there are no errors in the code.

14. Double-click the **Drawing_Cleanup** rule in the iLogic browser to open it.

When you tested changing the *Vise_Size* to **Medium** in the assembly, the left side drawing views overlapped the drawing border and the scale on the iso view was large for the sheet. Using a Drawing function, you can control the view position and size.

15. Place the cursor in the ElseIf statement for the **Medium Vise_Size**.

16. Expand the *Drawing* category in the *Snippets* area. Double-click **View Scale** to insert the snippet into the rule.

17. Add a second View Scale snippet on the next line.

View Names can be found in the Model Browser. Hover the cursor over a view in the drawing and its name will highlight in the Model Browser list.

18. Edit the two scaling functions to reflect which views are being changed and the scale value that should be used. Enter the expression, as shown in Figure 6–12.

19. Expand the *Drawing* category in the *Snippets* area. Double-click **View Set Center** to insert the snippet into the rule. Edit the name of the view that is being controlled in the first bracket and enter the X, Y values in the second set of brackets. Enter the expression, as shown in Figure 6–12.

20. Expand the *Drawing* category in the *Snippets* area. Double-click on View from Corner to insert the snippet into the rule. Edit the name of the view that is being controlled in the first bracket and enter the X, Y values, followed by the **SheetCorner.TopRight** parameter in the second set of brackets. This function is used to change the corner to which the view is being measured. Enter the expression, as shown in Figure 6–12.

```
If...Then...End If ▼ Keywords ▼ Operators ▼

'Controls view scale and positioning for Sheet 1
ActiveSheet = ThisDrawing.Sheet("Sheet:1")
modelName = IO.Path.GetFileName(ThisDrawing.ModelDocument.FullFileName)


If Parameter(modelname, "Vise_Size") = "Small" Then


ElseIf Parameter(modelname, "Vise_Size") = "Medium" Then
ActiveSheet.View("VIEW1").Scale = 1
ActiveSheet.View("VIEW4").Scale = 3/4
ActiveSheet.View("VIEW1").SetCenter(7, 6)
ActiveSheet.View("VIEW4").SetSpacingToCorner(.5, .5, SheetCorner.TopRight)


ElseIf Parameter(modelname, "Vise_Size") = "Large" Then


End If
```

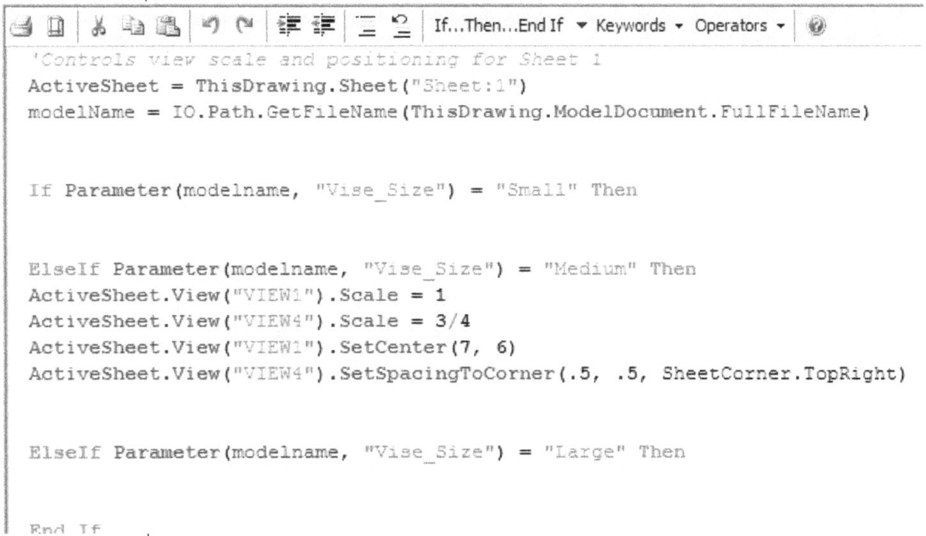

Figure 6–12

Determining the values used to reposition views involves trial and error. The values provided in this practice are simply one possible scenario.

21. Click **OK** to complete the rule. Ensure that the rule closes successfully to verify that there are no errors in the code.

22. Return to the **Vise.iam** model and change the **Vise_Size** parameter to **Medium**. Return to the drawing. The drawing updated in terms of the model size; however, the rule must be explicitly run to reposition views.

23. Right-click the **Drawing_Cleanup** rule in the iLogic browser and select **Run Rule**. The drawing should update as shown in Figure 6–13.

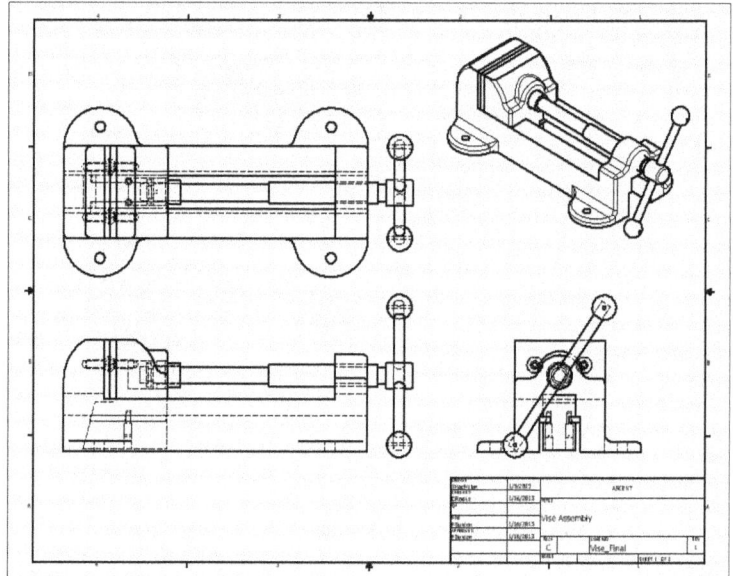

Figure 6–13

24. Return to the **Vise.iam** model and change the **Vise_Size** parameter back to **Small**. Return to the drawing. Note how the size of the model updates but the view positions and scales remain the same as those that were previously assigned for the Medium configuration.

25. Right-click on the **Drawing_Cleanup** rule in the iLogic browser and select **Run Rule**. The size and positions are still the same because no functions were added to the Small conditional scenario. Similar size and position functions must be added in all scenarios; otherwise the last assigned is maintained.

26. Double-click the **Drawing_Cleanup** rule in the iLogic browser to open it.

27. Copy and paste the four functions in the Medium condition that control the scale and location of the views into the Small condition. Edit the statements, as shown in Figure 6–14.

```
 If...Then...End If ▾  Keywords ▾  Operators ▾  ⓦ
'Controls view scale and positioning for Sheet 1
ActiveSheet = ThisDrawing.Sheet("Sheet:1")
modelName = IO.Path.GetFileName(ThisDrawing.ModelDocument.FullFileName)


If Parameter(modelname, "Vise_Size") = "Small" Then
ActiveSheet.View("VIEW1").Scale = 1
ActiveSheet.View("VIEW4").Scale = 1
ActiveSheet.View("VIEW1").SetCenter(6, 6)
ActiveSheet.View("VIEW4").SetSpacingToCorner(.5, .5, SheetCorner.TopRight)

ElseIf Parameter(modelname, "Vise_Size") = "Medium" Then
ActiveSheet.View("VIEW1").Scale = 1
ActiveSheet.View("VIEW4").Scale = 3/4
ActiveSheet.View("VIEW1").SetCenter(7, 6)
ActiveSheet.View("VIEW4").SetSpacingToCorner(.5, .5, SheetCorner.TopRight)

ElseIf Parameter(modelname, "Vise_Size") = "Large" Then
```

Figure 6–14

28. Once again, copy and paste the four functions in the Medium condition that control the scale and location of the views into the Large condition. Edit the statements, as shown in Figure 6–15.

```
If Parameter(modelname, "Vise_Size") = "Small" Then
ActiveSheet.View("VIEW1").Scale = 1
ActiveSheet.View("VIEW4").Scale = 1
ActiveSheet.View("VIEW1").SetCenter(6, 6)
ActiveSheet.View("VIEW4").SetSpacingToCorner(.5, .5, SheetCorner.TopRight)

ElseIf Parameter(modelname, "Vise_Size") = "Medium" Then
ActiveSheet.View("VIEW1").Scale = 1
ActiveSheet.View("VIEW4").Scale = 3/4
ActiveSheet.View("VIEW1").SetCenter(7, 6)
ActiveSheet.View("VIEW4").SetSpacingToCorner(.5, .5, SheetCorner.TopRight)

ElseIf Parameter(modelname, "Vise_Size") = "Large" Then
ActiveSheet.View("VIEW1").Scale = .8
ActiveSheet.View("VIEW4").Scale = 3/4
ActiveSheet.View("VIEW1").SetCenter(7, 6)
ActiveSheet.View("VIEW4").SetSpacingToCorner(.5, .5, SheetCorner.TopRight)
```

Figure 6–15

29. Test the Large configuration by changing the **Vise_Size** parameter in the assembly to **Large**. Return to the drawing. The size updates but the views overlap the edge.

30. Run the **Drawing_Cleanup** rule. The size of the model updates along with the newly assigned view positions and scales, as shown in Figure 6–16. The only issue in this drawing is the overlap of VIEW3 and VIEW4.

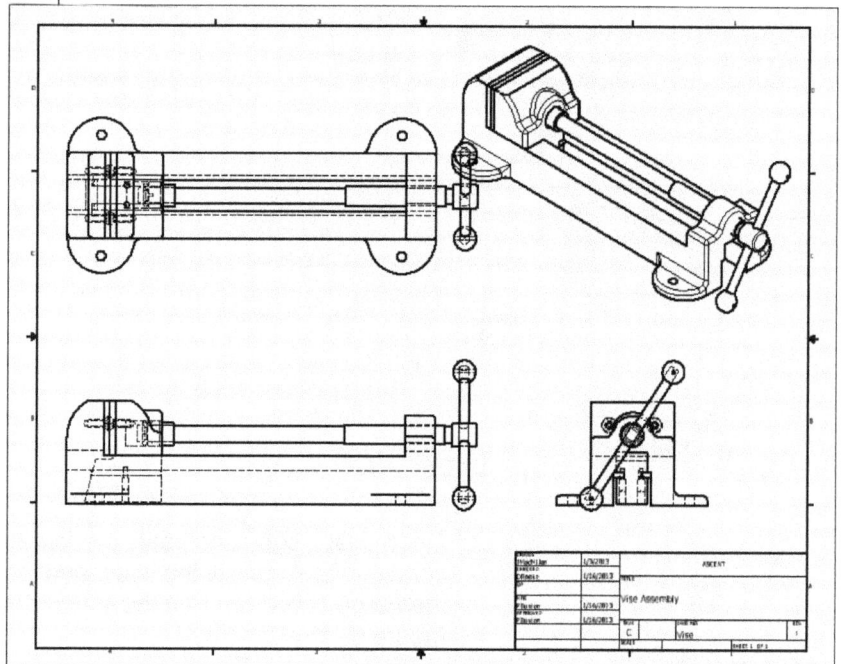

Figure 6–16

For the **Large Vise** configuration, the spacing between the views can be made a little larger to eliminate the overlap.

31. Double-click on the **Drawing_Cleanup** rule in the iLogic browser to open it.

32. Place the cursor on an empty line at the bottom of the functions listed in the Large conditional statement.

33. Expand the *Drawing* category in the *Snippets* area. Double-click on **Spacing to other View** to insert the snippet into the rule.

34. Add a second instance of the same snippet.

35. Edit the names of the views in the **Spacing to other View** functions and enter the distance values shown in Figure 6–17. The first view name will move (VIEW 2 and VIEW 3) and the second view name is the one being used as the measuring reference (VIEW1).

```
ActiveSheet.View("VIEW4").SetSpacingToCorner(.5, .5, SheetCorner.TopRight)

ElseIf Parameter(modelname, "Vise_Size") = "Medium" Then
ActiveSheet.View("VIEW1").Scale = 1
ActiveSheet.View("VIEW4").Scale = 3/4
ActiveSheet.View("VIEW1").SetCenter(7, 6)
ActiveSheet.View("VIEW4").SetSpacingToCorner(.5, .5, SheetCorner.TopRight)

ElseIf Parameter(modelname, "Vise_Size") = "Large" Then
ActiveSheet.View("VIEW1").Scale = .8
ActiveSheet.View("VIEW4").Scale = 3/4
ActiveSheet.View("VIEW1").SetCenter(7, 6)
ActiveSheet.View("VIEW4").SetSpacingToCorner(.5, .5, SheetCorner.TopRight)
ActiveSheet.View("VIEW2").SpacingBetween("VIEW1") = 2
ActiveSheet.View("VIEW3").SpacingBetween("VIEW1") = .5

End If
```

Figure 6–17

Similar to how you are required to enter the scale and positions in each condition, you are required to do the same with the spacing.

36. Copy and paste the two **Spacing to other View** functions into the Small and Medium conditional statements and edit them, as shown in Figure 6–18.

```
If...Then...End If ▾ Keywords ▾ Operators ▾

'Controls view scale and positioning for Sheet 1
ActiveSheet = ThisDrawing.Sheet("Sheet:1")
modelName = IO.Path.GetFileName(ThisDrawing.ModelDocument.FullFileName)


If Parameter(modelname, "Vise_Size") = "Small" Then
ActiveSheet.View("VIEW1").Scale = 1
ActiveSheet.View("VIEW4").Scale = 1
ActiveSheet.View("VIEW1").SetCenter(6, 6)
ActiveSheet.View("VIEW4").SetSpacingToCorner(.5, .5, SheetCorner.TopRight)
ActiveSheet.View("VIEW2").SpacingBetween("VIEW1") = 2
ActiveSheet.View("VIEW3").SpacingBetween("VIEW1") = 1

ElseIf Parameter(modelname, "Vise_Size") = "Medium" Then
ActiveSheet.View("VIEW1").Scale = 1
ActiveSheet.View("VIEW4").Scale = 3/4
ActiveSheet.View("VIEW1").SetCenter(7, 6)
ActiveSheet.View("VIEW4").SetSpacingToCorner(.5, .5, SheetCorner.TopRight)
ActiveSheet.View("VIEW2").SpacingBetween("VIEW1") = 2
ActiveSheet.View("VIEW3").SpacingBetween("VIEW1") = .5

ElseIf Parameter(modelname, "Vise_Size") = "Large" Then
ActiveSheet.View("VIEW1").Scale = .8
ActiveSheet.View("VIEW4").Scale = 3/4
ActiveSheet.View("VIEW1").SetCenter(7, 6)
ActiveSheet.View("VIEW4").SetSpacingToCorner(.5, .5, SheetCorner.TopRight)
ActiveSheet.View("VIEW2").SpacingBetween("VIEW1") = 2
ActiveSheet.View("VIEW3").SpacingBetween("VIEW1") = .5
```

Figure 6–18

37. Click **OK** to complete the rule. Ensure that the rule closes successfully to verify that there are no errors in the code.

38. Return to the **Vise.iam** model and change the **Vise_Size** parameter to another option. Return to the drawing and rerun the rule. Ensure that the positioning and scaling for all three configurations works correctly.

Task 3 - Edit the drawing rule to incorporate a change of title block and border for the Large Vise configuration.

In this task you will further customize the drawing sheet by swapping the title block and border of the sheet whenever the Large Vise configuration is specified. The new title block and border does not exist in this current drawing. It will be pulled from a reference drawing, when required. To identify the reference drawing you must include the code in the rule.

1. Double-click on the **Drawing_Cleanup** rule to edit it.

2. Place the cursor on the fourth line (below the modelName entry). Expand the *Drawing* category in the *Snippets* area. Double-click on the **ResourceFileName** to insert the snippet into the rule. Edit the drawing name to **Generic Template.idw**, as shown in Figure 6–19. This identifies the resource drawing that the title block and border can be pulled from, if not found in the current drawing.

3. Expand the *Drawing* category in the *Snippets* area. Double-click on the **KeepExtraResources** to insert the snippet into the rule, as shown in Figure 6–19.

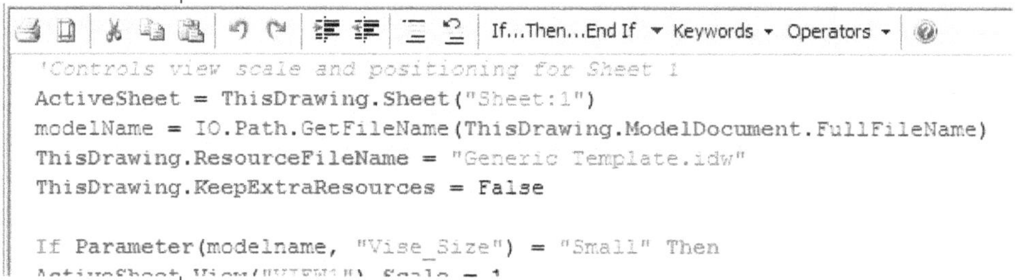

```
'Controls view scale and positioning for Sheet 1
ActiveSheet = ThisDrawing.Sheet("Sheet:1")
modelName = IO.Path.GetFileName(ThisDrawing.ModelDocument.FullFileName)
ThisDrawing.ResourceFileName = "Generic Template.idw"
ThisDrawing.KeepExtraResources = False

If Parameter(modelname, "Vise_Size") = "Small" Then
ActiveSheet.View("VIEW1").Scale = 1
```

Figure 6–19

By including the **KeepExtraResources** function in the rule, any resources that are pulled from the resource drawing are only temporarily stored in the current drawing. If they are not being used, they are automatically deleted. If this value is set to *True*, the resource would be saved in the current drawing. Setting this to *False* is valuable if you have a global template that you are pulling from, if changes are made to it, the drawing pulls the updated template when used again.

Similar to positioning and scaling, the title block and border assignments must be made in each configuration, otherwise, the last assigned will be maintained.

4. In the Small configuration's conditional statement, add the **Change Title Block** and **Change Border** functions at the top of the condition. Note in the Model browser that the Default Border and ANSI - Large title blocks are being used. Edit the values in the rule, as shown in Figure 6–20, to maintain these resources for this configuration.

```
'Controls view scale and positioning for Sheet 1
ActiveSheet = ThisDrawing.Sheet("Sheet:1")
modelName = IO.Path.GetFileName(ThisDrawing.ModelDocument.FullFileName)
ThisDrawing.ResourceFileName = "Generic Template.idw"
ThisDrawing.KeepExtraResources = False


If Parameter(modelname, "Vise_Size") = "Small" Then
ActiveSheet.TitleBlock = "ANSI - Large"
ActiveSheet.Border = "Default Border"
ActiveSheet.View("VIEW1").Scale = 1
ActiveSheet.View("VIEW4").Scale = 1
ActiveSheet.View("VIEW1").SetCenter(6, 6)
```

Figure 6–20

5. Copy the **Change Title Block** and **Change Border** functions to the Medium configuration's conditional statement, as shown in Figure 6–21. In this configuration, the title block and border also remain the same.

```
ActiveSheet.View("VIEW1").SetSpacingToCorner(.5, .5, SheetCorner.TopRight)
ActiveSheet.View("VIEW2").SpacingBetween("VIEW1") = 2
ActiveSheet.View("VIEW3").SpacingBetween("VIEW1") = 1

ElseIf Parameter(modelname, "Vise_Size") = "Medium" Then
ActiveSheet.TitleBlock = "ANSI - Large"
ActiveSheet.Border = "Default Border"
ActiveSheet.View("VIEW1").Scale = 1
ActiveSheet.View("VIEW4").Scale = 3/4
ActiveSheet.View("VIEW1").SetCenter(7, 6)
ActiveSheet.View("VIEW4").SetSpacingToCorner(.5, .5, SheetCorner.TopRight)
ActiveSheet.View("VIEW2").SpacingBetween("VIEW1") = 2
ActiveSheet.View("VIEW3").SpacingBetween("VIEW1") = .5
```

Figure 6–21

6. Copy the **Change Title Block** and **Change Border** functions to the Large configuration's conditional statement. For this configuration a new title block and border are to be assigned from the resource drawing.

7. Edit the **Change Title Block** and **Change Border** functions to define the new title block and border, as shown in Figure 6–22

```
ActiveSheet.View("VIEW1").SetCenter(7, 6)
ActiveSheet.View("VIEW4").SetSpacingToCorner(.5, .5, SheetCorner.TopRight)
ActiveSheet.View("VIEW2").SpacingBetween("VIEW1") = 2
ActiveSheet.View("VIEW3").SpacingBetween("VIEW1") = .5

ElseIf Parameter(modelname, "Vise_Size") = "Large" Then
ActiveSheet.TitleBlock = "Main Title Block"
ActiveSheet.Border = "C-Border"
ActiveSheet.View("VIEW1").Scale = .8
ActiveSheet.View("VIEW4").Scale = 3/4
ActiveSheet.View("VIEW1").SetCenter(7, 6)
ActiveSheet.View("VIEW4").SetSpacingToCorner(.5, .5, SheetCorner.TopRight)
ActiveSheet.View("VIEW2").SpacingBetween("VIEW1") = 2
ActiveSheet.View("VIEW3").SpacingBetween("VIEW1") = .5

End If
```

Figure 6–22

8. Click **OK** to complete the rule. Ensure that the rule closes successfully to verify that there are no errors in the code.

9. Return to the **Vise.iam** model and change the **Vise_Size** parameter to **Large**, if not already set. Return to the drawing and rerun the rule. Note the new title block and border.

10. Test returning to the Small or Medium configurations and ensure that the title block and border return to its original when you run the drawing rule. Also note how the Main Title Block and C-Border resources are no longer in the current drawing's Drawing Resources.

11. Save the drawing and close all files.

*A completed rule has been included in the **Vise_Final.idw** file in the practice files for you to review.*

Later, you will learn about triggers and how they can be used to automatically initiate a rule.

Chapter Review Questions

1. Which of the following can be controlled using an iLogic drawing rule? (Select all that apply.)

   a. Sheet size based on specific view sizes.

   b. View placement relative to a corner of the drawing.

   c. View placement relative to another view.

   d. View suppression.

   e. Dimensions associated with a view.

2. Which of the following functions can be used to change the size of a drawing sheet?

   a. ActiveSheet.View("VIEW1").Width

   b. ActiveSheet.ChangeSize("A3")

   c. ActiveSheet.TitleBlock = "<ANSI - Large>"

   d. ActiveSheet.Border = "<Default Border>"

3. After the following functions are run in a drawing rule, what is the value of the ActiveSheet variable?

   ```
   ActiveSheet = ThisDrawing.Sheet("Isometrics:1")
   ActiveSheet.TitleBlock = "<Custom>"
   ActiveSheet.View("4")
   ThisDrawing.ResourceFileName = "<DrawingResources.idw>"
   ```

   a. Isometrics:1

   b. Custom

   c. 4

   d. DrawingResources.idw

4. Which of the following functions should you use if changes to a global title block are frequent and you only want to change one title block versus making changes independently in all drawings that use the title block?

   a. ThisDrawing.KeepExtraResources = True

   b. ThisDrawing.KeepExtraResources = False

5. Which of the following lines of code should you use if you want to change the scale value of a view to 1:2?

   a. ActiveSheet.View("VIEW1").Scale

   b. ActiveSheet.View("VIEW1").ScaleString

6. Which of the following lines of code should you use if you want to change the view position relative to another view?

   a. ActiveSheet.View("VIEW1").SetCenter(1, 1)

   b. ActiveSheet.View("VIEW1").SetSpacingToCorner(1, 1, SheetCorner.TopLeft)

   c. ActiveSheet.View("2").SpacingBetween("1") = 1

   d. ActiveSheet.View("VIEW1").Width

   e. ActiveSheet.View("VIEW1").Height

7. Which of the following functions enables you to suppress a VIEW2 in a drawing?

   a. ActiveSheet.View("VIEW2").View.Suppressed = True

   b. ActiveSheet.View("VIEW2").View.Suppressed = False

Command Summary

Button	Command	Location
	Add Rule	• **Ribbon:** *Manage* tab>iLogic panel

Rule Triggering and Form Creation

Event Trigger functionality enables you to associate an iLogic rule that exists in a part, assembly, or drawing model with a specific Autodesk® Inventor® event. When the Autodesk Inventor event is activated, the assigned rule is executed. As an alternative to associating a rule with an event, you can also establish an iTrigger that involves the use of an Autodesk Inventor parameter. In either situation, triggering enables you to force iLogic rules to execute instead of manually executing them. Forms provide a customizable interface that enables easy interaction with parameters, rules, and iProperties.

Learning Objectives in this Chapter

- Associate rules to specific Autodesk Inventor actions so that they are executed when the action occurs.
- Incorporate the use of iTriggers so that they can be used to execute iLogic rules.
- Create a custom form that enables the entry of parameter and iProperty values and that can execute iLogic rules.

iLogic Workflow

Figure 7–1 shows the overall suggested workflow for the iLogic tools in the Autodesk Inventor software. The horizontal line at the top represents the high-level workflow and each of their sub-steps are detailed vertically below them. The highlighted column represents the content discussed in the current chapter.

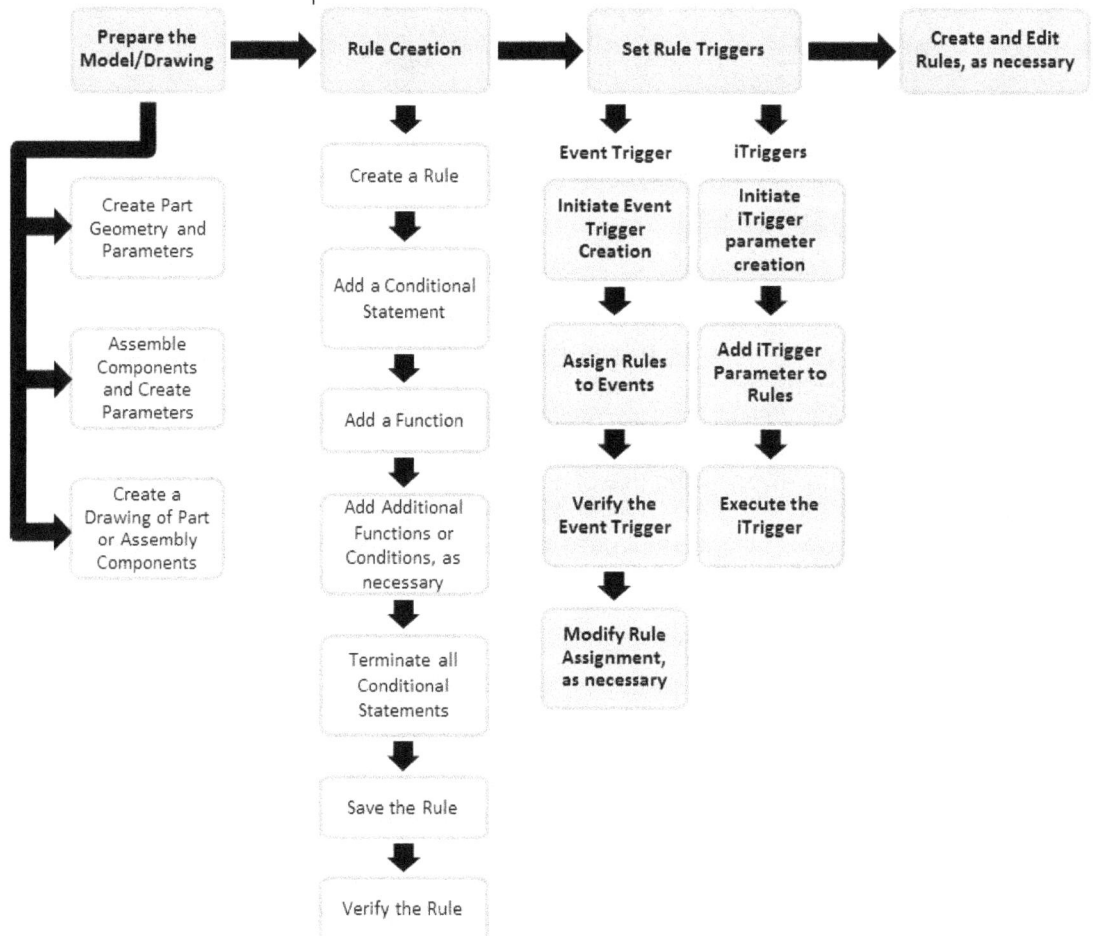

```
Prepare the        →    Rule Creation    →    Set Rule Triggers    →    Create and Edit
Model/Drawing                                                             Rules, as necessary
```

Prepare the Model/Drawing	Rule Creation	Set Rule Triggers		Create and Edit Rules, as necessary

Event Trigger | iTriggers

Create Part Geometry and Parameters

Create a Rule

Initiate Event Trigger Creation | Initiate iTrigger parameter creation

Add a Conditional Statement

Assemble Components and Create Parameters

Assign Rules to Events | Add iTrigger Parameter to Rules

Add a Function

Create a Drawing of Part or Assembly Components

Add Additional Functions or Conditions, as necessary

Verify the Event Trigger | Execute the iTrigger

Terminate all Conditional Statements

Modify Rule Assignment, as necessary

Save the Rule

Verify the Rule

Figure 7–1

7.1 Event Triggers

Once rules are created in a part or assembly model and you have verified that they work as expected in the model, you can define event triggers. Event triggers enable you to specifically define when a rule is triggered based on the occurrence of a standard Autodesk Inventor event. iLogic provides a list of triggers to which the established rules can be assigned.

General Steps

Use the following general steps to establish an event trigger for an iLogic rule:

1. Initiate event trigger creation.
2. Assign rules to events.
3. Verify the event trigger.
4. Modify rule assignment, as necessary.

Figure 7–2 highlights the steps graphically. Additional in-depth information is included on each of the rule creation steps in the remainder of this section.

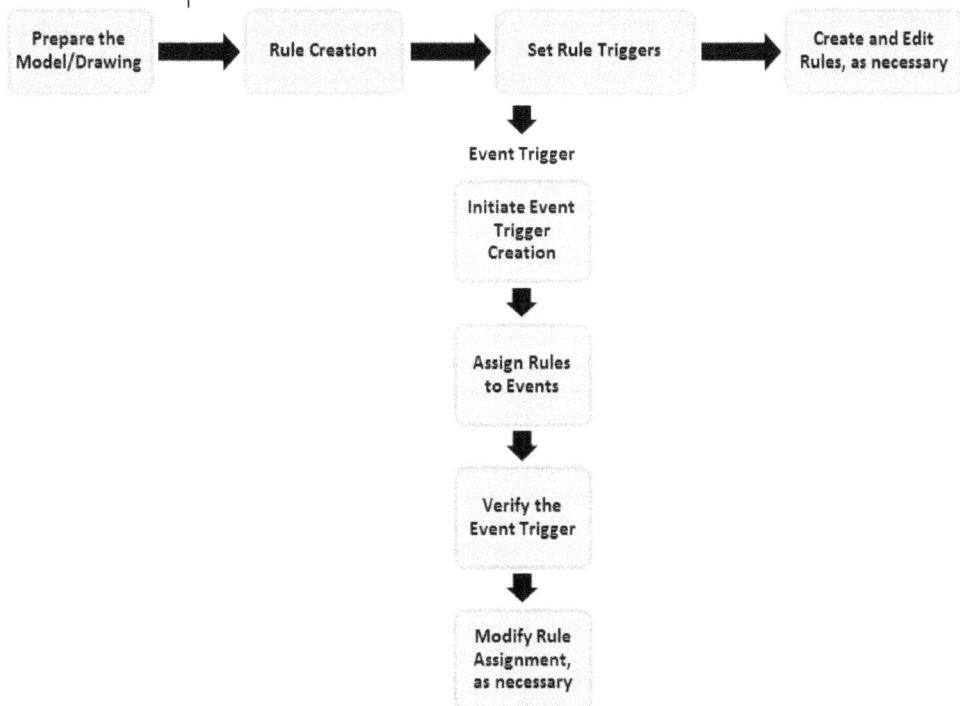

Prepare the Model/Drawing → Rule Creation → Set Rule Triggers → Create and Edit Rules, as necessary

Event Trigger

Initiate Event Trigger Creation

Assign Rules to Events

Verify the Event Trigger

Modify Rule Assignment, as necessary

Figure 7–2

Step 1 - Initiate event trigger creation.

In the *Manage* tab>iLogic panel, click ![Event Triggers icon] (Event Triggers) to create triggers. The Rules Triggered by Events dialog box opens, as shown in Figure 7–3. The list of events vary slightly depending on whether a part, assembly, or drawing is active. The available events provide you with options to link a specific Autodesk Inventor action with the execution of an iLogic rule.

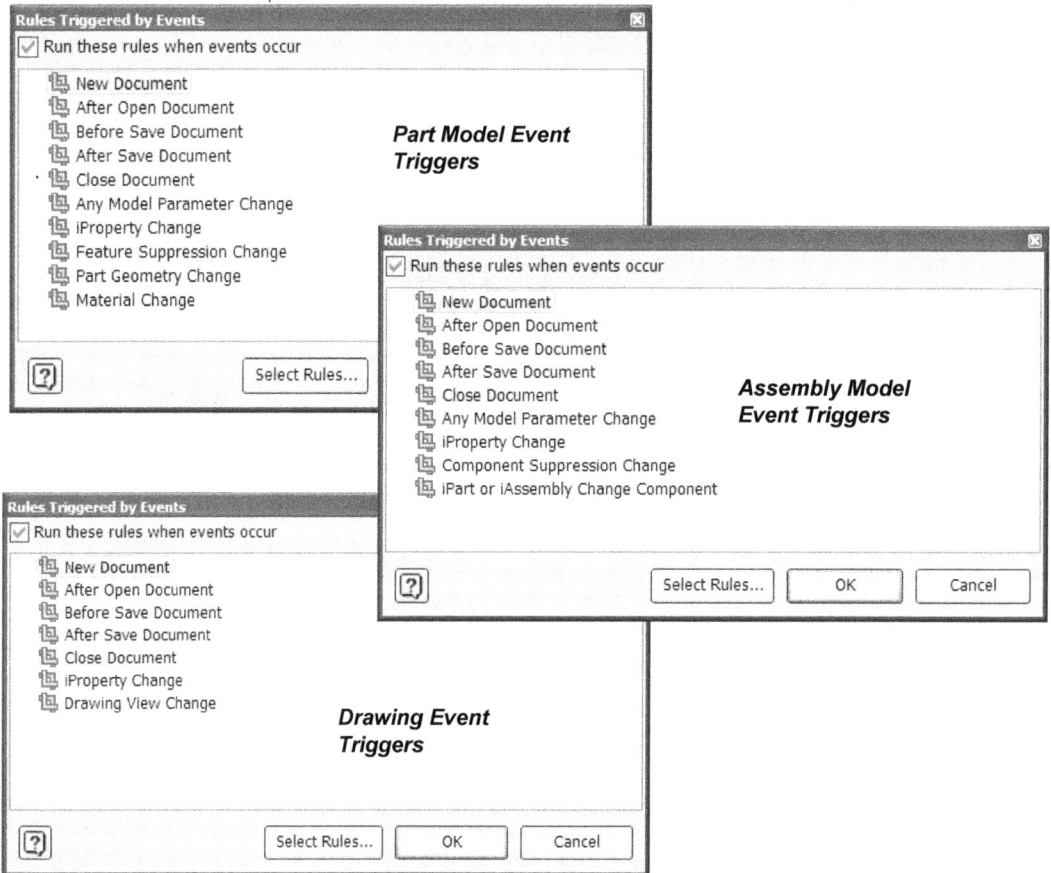

Rules Triggered by Events

☑ Run these rules when events occur

- New Document
- After Open Document
- Before Save Document
- After Save Document
- Close Document
- Any Model Parameter Change
- iProperty Change
- Feature Suppression Change
- Part Geometry Change
- Material Change

Part Model Event Triggers

[?] Select Rules...

Rules Triggered by Events

☑ Run these rules when events occur

- New Document
- After Open Document
- Before Save Document
- After Save Document
- Close Document
- Any Model Parameter Change
- iProperty Change
- Component Suppression Change
- iPart or iAssembly Change Component

Assembly Model Event Triggers

[?] Select Rules... OK Cancel

Rules Triggered by Events

☑ Run these rules when events occur

- New Document
- After Open Document
- Before Save Document
- After Save Document
- Close Document
- iProperty Change
- Drawing View Change

Drawing Event Triggers

[?] Select Rules... OK Cancel

Figure 7–3

Consider the use of iTrigger if you cannot control rule execution with any of the predefined event trigger options.

The Event Triggers that are available are predefined. You cannot create custom event triggers. The events that can be used as triggers to initiate a rule are described as follows:

Event Trigger	Description
New Document	A rule is triggered when a new file is created.
After Open Document	A rule is triggered when a file is opened.
Before Save Document	A rule is triggered before a file is saved.
After Save Document	A rule is triggered after a file is saved.
Close Document	A rule is triggered before a file is closed.
Any Model Parameter Change	A rule is triggered when any model parameter is changed.
iProperty Change	A rule is triggered when an iProperty is changed.
Feature Suppression Change (Part only)	A rule is triggered when a feature in a part file is suppressed or unsuppressed.
Part Geometry Change (Part only)	A rule is triggered when a model geometry is changed.
Material Change (Part only)	A rule is triggered when a model's material is changed.
Component Suppression Change (Asm only)	A rule is triggered when a component of an assembly is suppressed or unsuppressed.
iPart or iAssembly Change Component (Asm only)	A rule is triggered when a component of an iPart or iAssembly is changed.
Drawing View Change (Drawing only)	A rule is triggered when a drawing view is changed.

Step 2 - Assign rules to events.

To assign an existing rule to an event, highlight the event that is to be used and click **Select Rules**. Alternatively, you can double-click the Event name or right-click and select **Select Rules**. The Choose Rules for <Event Name> dialog box opens. Figure 7–4 shows the Choose Rules for Material Change dialog box which opens when the Material Change event trigger is edited.

External rules that are created outside of the Autodesk Inventor software can also be assigned to an event trigger. If they exist, they are displayed in the External rules area of the dialog box and can be selected in the same way as the rules that exist in the document.

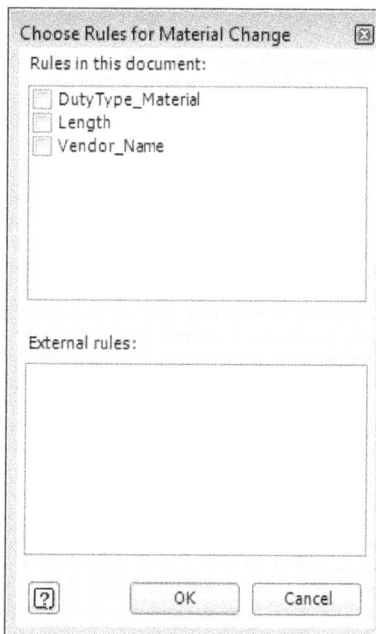

Figure 7–4

The complete list of rules that exist in the document display in the *Rules in this document* area at the top of the dialog box. Select the rule that is to be assigned to the trigger and click **OK**. The Rules Triggered by Events dialog box updates and the rule that is assigned to the event is listed beneath it, as shown in Figure 7–5.

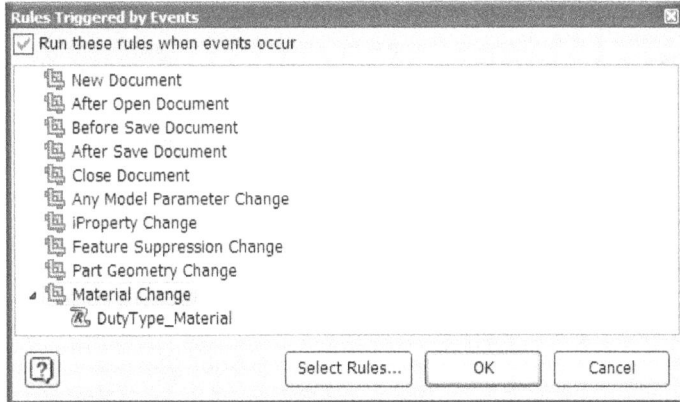

Figure 7–5

- If multiple rules are assigned to an event, they can be dragged in the list to reorder the order in which the rules are executed when the event occurs.

- To globally disable all event triggers that have been assigned from executing their assigned rules, clear the **Run these rules when events occur** option at the top of the dialog box.

To complete the event trigger, click **OK**.

Step 3 - Verify the event trigger.

To ensure that a rule is executed when its associated Autodesk Inventor event is executed, you must explicitly execute the event. This enables you to verify the correct execution of the rule.

Step 4 - Modify rule assignment, as necessary.

To make a change to events to which rules are assigned, in the *Manage* tab>iLogic panel, click (Event Triggers). The dialog box opens similar to that shown in Figure 7–6.

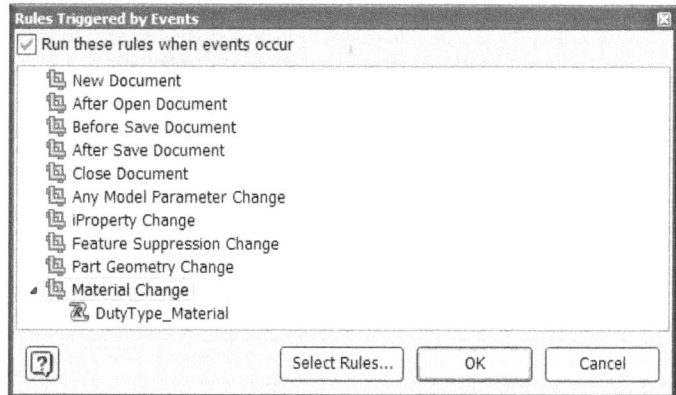

Figure 7–6

- To edit an existing rule to an event, highlight the event that is to be used and click **Select Rules**. Alternatively, you can double-click on the Event name or right-click and select **Select Rules**. In the Choose Rules <Event Name> dialog box, clear checkboxes to remove rules from the event or select checkboxes to add new rules to the event trigger.

7.2 iTriggers

The use of an iTrigger provides an alternative to associating rules with events using the Event Trigger functionality.

iTrigger functionality creates an Autodesk Inventor user parameter in the model that can be subsequently referenced in any iLogic rules in the document to force rule execution. When the **iTrigger** command is activated, any rule(s) that have the **iTrigger** parameter in them, are automatically executed.

General Steps

Use the following general steps to create an iTrigger for an iLogic rule:

1. Initiate **iTrigger** parameter creation.
2. Add **iTrigger** parameter to rules.
3. Execute the iTrigger.

Figure 7–7 highlights the steps graphically. Additional in-depth information is included on each of the rule creation steps in the remainder of this section.

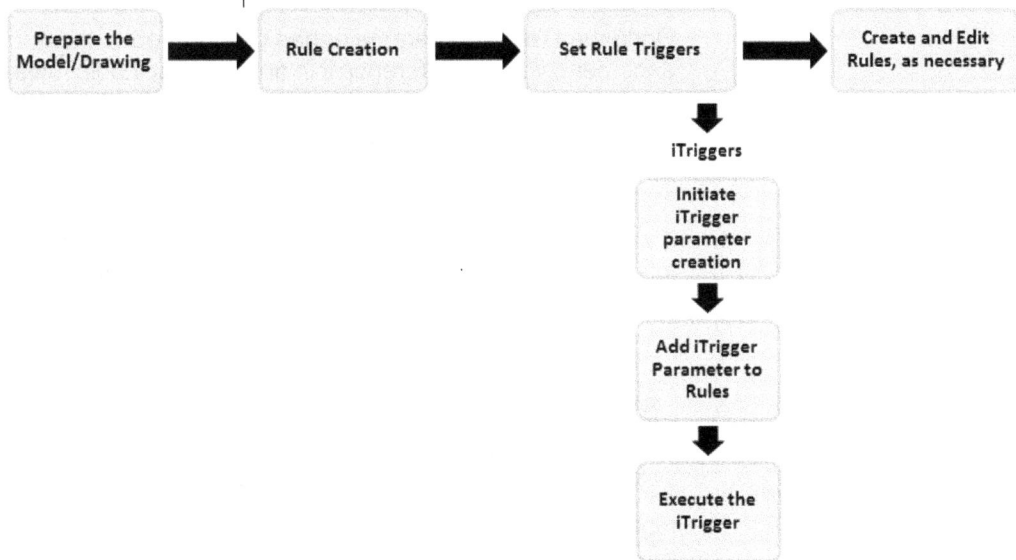

Prepare the Model/Drawing → Rule Creation → Set Rule Triggers → Create and Edit Rules, as necessary

iTriggers

Initiate iTrigger parameter creation

Add iTrigger Parameter to Rules

Execute the iTrigger

Figure 7–7

Step 1 - Initiate iTrigger parameter creation.

In the *Manage* tab>iLogic panel, click ✥ (iTrigger) to create the **iTrigger** parameter in the current document. Once created, you can verify that it was created in the Parameters dialog box. The **iTrigger0** parameter should be listed in the **User Parameter** node of the dialog box with a value of **1**, as shown in Figure 7–8.

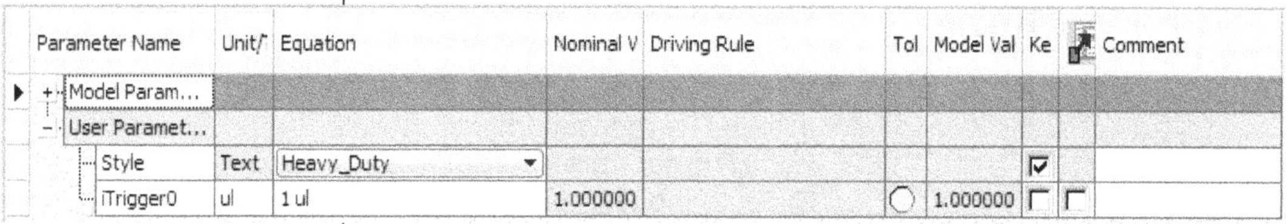

Parameter Name	Unit/	Equation	Nominal V	Driving Rule	Tol	Model Val	Ke	🖉 Comment
▶ + Model Param...								
− User Paramet...								
Style	Text	Heavy_Duty ▼					☑	
iTrigger0	ul	1 ul	1.000000		○	1.000000	☐ ☐	

Figure 7–8

Step 2 - Add iTrigger parameter to rules.

Once the **iTrigger0** parameter has been added to the current document, you can reference it in any existing iLogic rules or you can create a new rule that references it. In either situation, the syntax that is used in the rule is as follows:

> trigger = iTrigger0

The iTrigger0 reference in the rule can be added in any location in a rule, but adding it at the top of the rule helps you to locate the line of code when reviewing an iLogic rule.

Step 3 - Execute the iTrigger.

Each time ✥ (iTrigger) is executed after the parameter is created, the equation value sequentially increases. This has no impact on the model. It can be used to evaluate how many times the rule has been run.

With the **iTrigger0** parameter referenced in a rule, in the *Manage* tab>iLogic panel, click ✥ (iTrigger) to automatically execute any rule that contains the iTrigger0 reference.

- Multiple rules might be executed if the **iTrigger0** parameter exists in multiple rules.

- Consider the order in which the rules are listed in the iLogic browser if the order of execution is important in the document.

7.3 Forms

The use of forms in the model enables you to design a custom interface that can be used to enter only the required information. You are not required to navigate to the Parameter or iProperties dialog boxes to make changes. All the required data is presented in the form, similar to that shown in Figure 7–9.

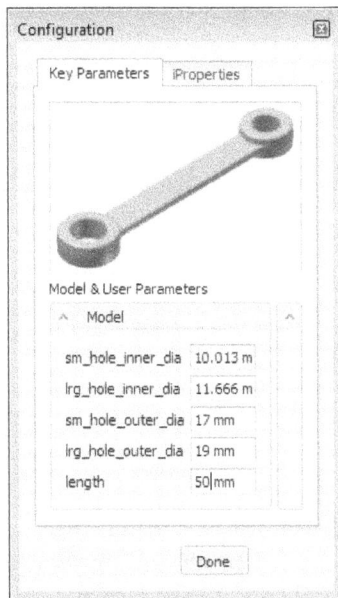

Figure 7–9

General Steps

Use the following general steps to create a form:

1. Start the creation of the form.
2. Customize the form layout.
3. Add controls to the form.
4. Customize the Properties of the form components.
5. Complete form Creation.
6. Verify the Form.
7. Edit the form, as required.

Step 1 - Start the creation of the form.

In the *Manage* tab>expand the iLogic panel, and click ⬚ (Add Form). The Add Form dialog box opens as shown in Figure 7–10.

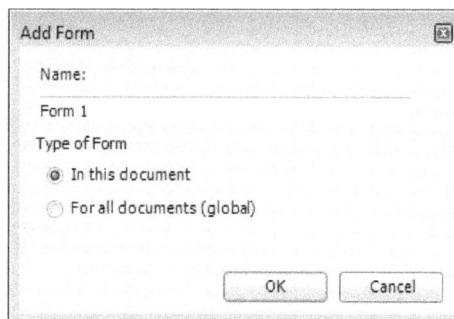

Add Form	☒
Name:	
Form 1	
Type of Form	
⦿ In this document	
○ For all documents (global)	
	OK Cancel

Figure 7–10

Enter a descriptive name for the form and select whether the form is for the current document (**In this document**) or is for global use (**For all documents (global)**).

- Forms created locally can be copied to other documents, if required.

Global Forms are commonly used for drawing templates.

- The **For all documents (global)** forms are saved to the *Design Data* folder for use by all documents.

  - If a change is required in the form it can be made in any file, versus editing the same form in multiple documents.
  - Global forms are stored in the *C:\Users\Public\Public Documents\Autodesk\Inventor 2017\DesignData\ iLogic\UI* directory. You can access this directory, by right-clicking the background of the *Global Forms* tab in the iLogic browser and select **Open Containing Folder**.

Click **OK**. Once the form is created the Form Editor dialog boxes opens as shown in Figure 7–11.

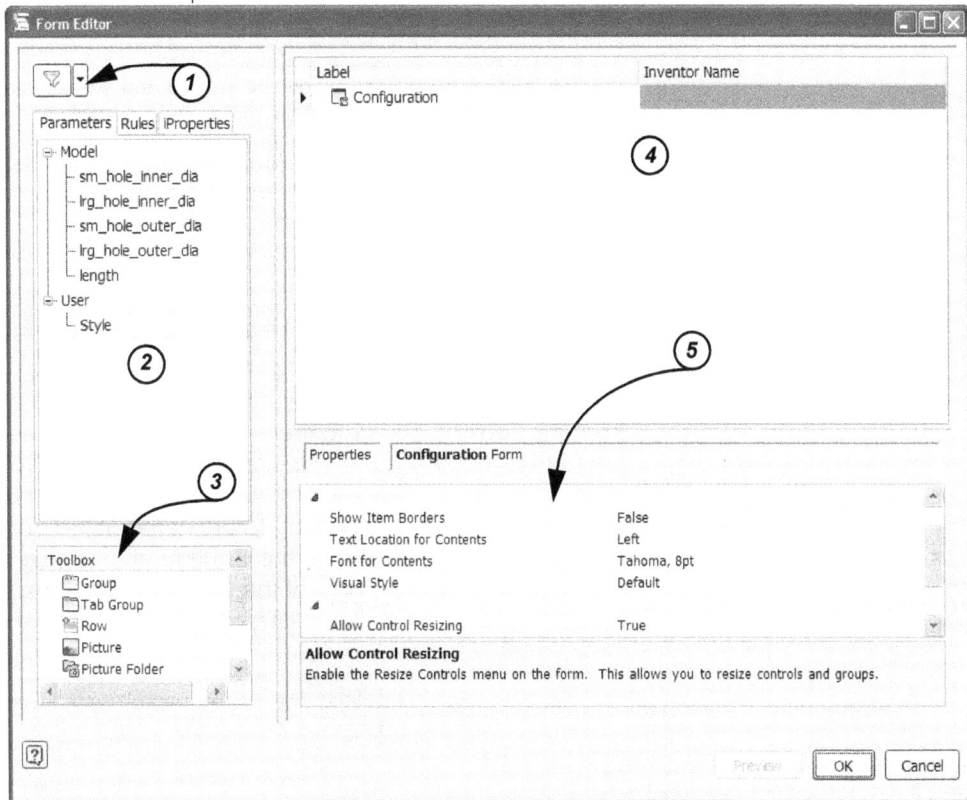

Figure 7–11

1. Filter	2. Tabs
3. Toolbox	4. Design Tree
5. Properties	

The areas of the Form Editor dialog box are described as follows:

Area	Description
1. Filter	Enables you to filter the list of parameters displayed in the *Parameters* tab. You can limit the list of displayed parameters to: • those identified as Key (**Key**). • those that have been renamed (**Renamed**). • you can show the entire list of all user and model parameters (**All**) in the model.

2. Tabs	The Tabs area provides a list of all Parameters, Rules, and iProperties in the current model.
3. Toolbox	Provides components that are used to customize the layout of a form.
4. Design Tree	Provides the overall working area where elements are dragged to create the overall form layout.
5. Properties	Enables you to customize the properties of selected elements in the Design Tree. The options available depend on the item selected.

Step 2 - Customize the form layout.

The Form Editor dialog box enables you to drag the elements from the Toolbox and drop them onto the Design Tree to create the form layout. As you are designing the layout, a Preview of the form displays in the dialog box adjacent to the Form Editor. This dialog box is named according to the name of the Form that is being created. For example, the Configuration dialog box shown in Figure 7–12 is a preview of the Configuration form before the layout is added to the Design Tree.

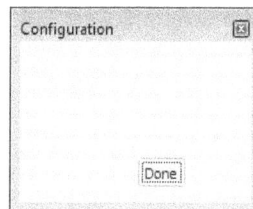

Figure 7–12

How To: Add any of the Layout Components from the Toolbox to the Design Tree

1. Select a layout component in the *Toolbox* area.
2. While pressing and holding the mouse button on the layout component in the *Toolbox* area, drag the component to the Design Tree.
3. Drop the layout component in the Design Tree at the desired position.
   - The Form preview dialog box updates as each component is added.

4. Double-click on any component in the Design Tree and rename it, as required, to further customize the names of Groups or Tabs.
5. The location of any of the layout components in the Design Tree can be changed by selecting and dragging it to a new position.
6. Continue to customize the overall layout of the form.

The layout components that are available in the toolbox are described as follows:

Component	Description
(Group)	Enables you to group items on the form. Groups are boxed and are collapsible, as shown for the Key Parameters Group in the following image. 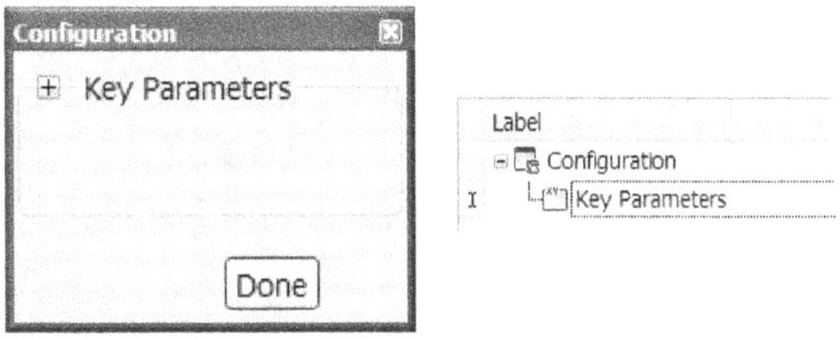
(Tab Group)	Enables you to create a tab on the form, as shown for the Tab 1 and Tab 2 in the following image. 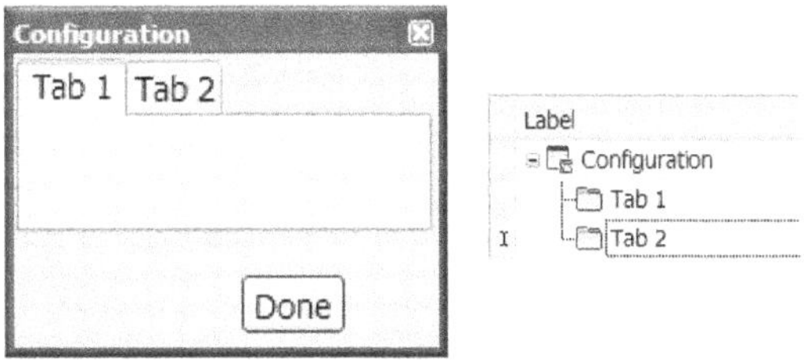
(Row)	Enables you to create a horizontal row on the form to organize controls. Controls are organized vertically by default. In the following image two groups have been added horizontally by nesting them in a Row component. 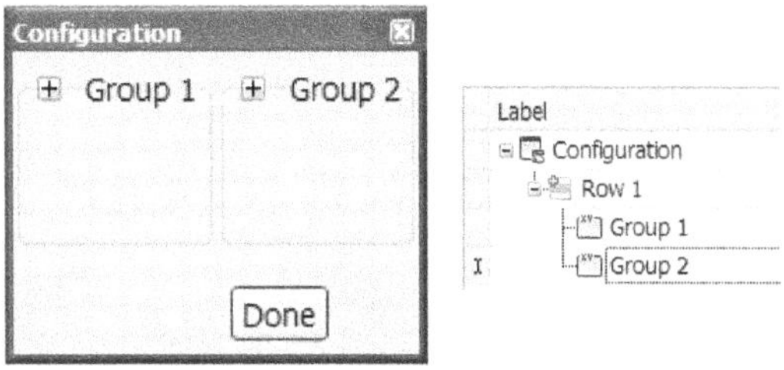

(Picture)	Enables you to add an informational picture to the form. This component only adds the component to the layout; you must define the image properties to insert the image file. 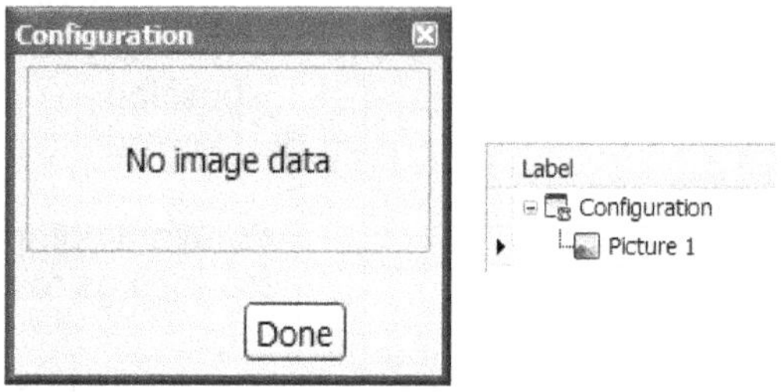

(Picture Folder)	Enables you to add a folder for alternate pictures.

(Empty Space)	Enables you to add a blank space between components on the form.

A (Label)	Enables you to add a text label on the form, as shown for Label 1 and Label 2 in the following image. 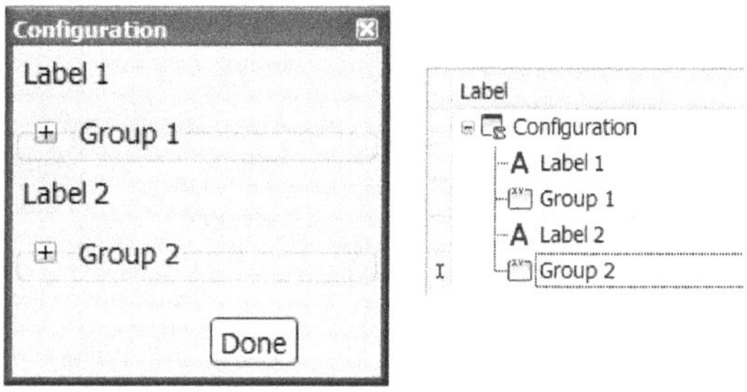

(Splitter)	Enables you to add a resizable splitter bar on the form. In the following image a splitter bar has been added after Group 1 in the layout which enables group one to be resized horizontally. Without a splitter bar, Group 2 cannot be resized horizontally. 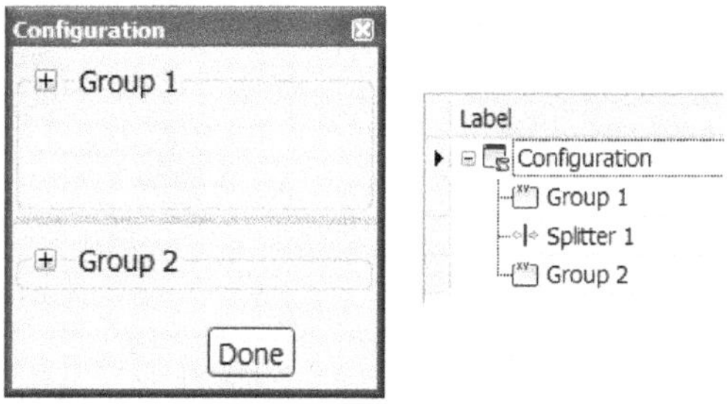

The form layout (shown in Figure 7–13) shows an example of a designed layout.

Once a component has been renamed refer to the thumbnail image adjacent to the component name to identify its type.

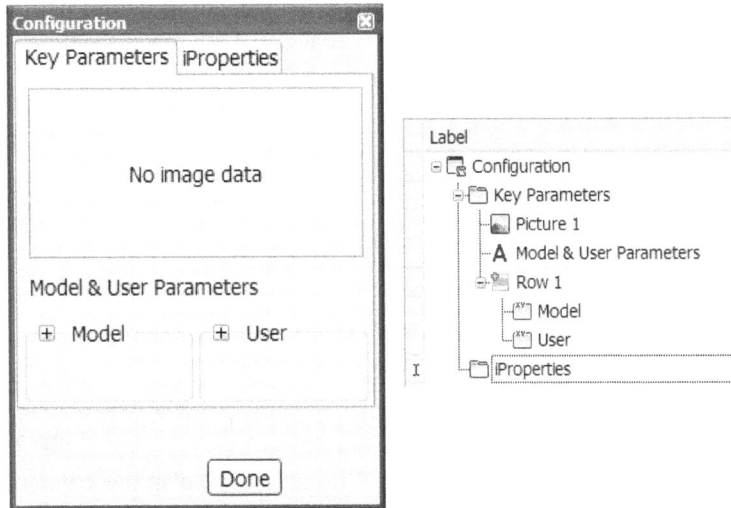

Figure 7–13

In this form, two tabs (⬜) have been added. Only the *Key Parameters* tab has additional layout components added to it. In the *Key Parameters* tab, a picture (🖼️) has been added to the top of the layout, followed by a row (⬚). The row enables the form to list any nested components horizontally, as is the case for the two groups (⬜) that have been added.

Step 3 - Add controls to the form.

Once the form layout has been defined, you must add controls to the components in the form. Controls are either model or user parameters, iLogic rules, or iProperties from the model that can be modified or executed to affect change in the model.

How To: Add Controls to a Layout Component in the Design Tree

1. Select the tab from which you want to select a control. The available tabs are *Parameters*, *Rules*, and *iProperties*, as shown in Figure 7–14.

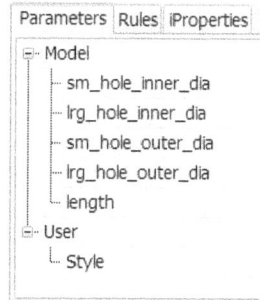

Figure 7–14

*If the Parameters tab is selected, consider using the Filter controls above the tabs to refine the list of parameters that are displayed. Filtering refines the list to display only those parameters that are required (**Key**, **Renamed**, or **All**)*

2. Select a control in the selected tab's list.
   - You can select individual controls or you can select multiple controls by using <Shift>.

3. While pressing and holding the mouse button on a selected control(s), drag onto the required component in the Design Tree.

   - For example, in Figure 7–15, five model parameters are selected in the *Parameters* tab and dragged onto the Model Group (^{xv}).

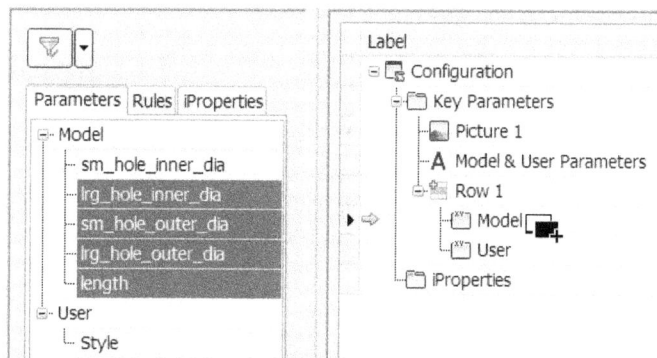

Figure 7–15

4. Release the mouse button to drop the controls into the component in the Design Tree.

- The Design Tree and the Form preview dialog boxes update as each control is added. For example, Figure 7–16 shows the resulting Design Tree Form preview dialog boxes after five model parameters were selected in the *Parameters* tab and dragged onto the Model Group ($\boxed{xv}$).

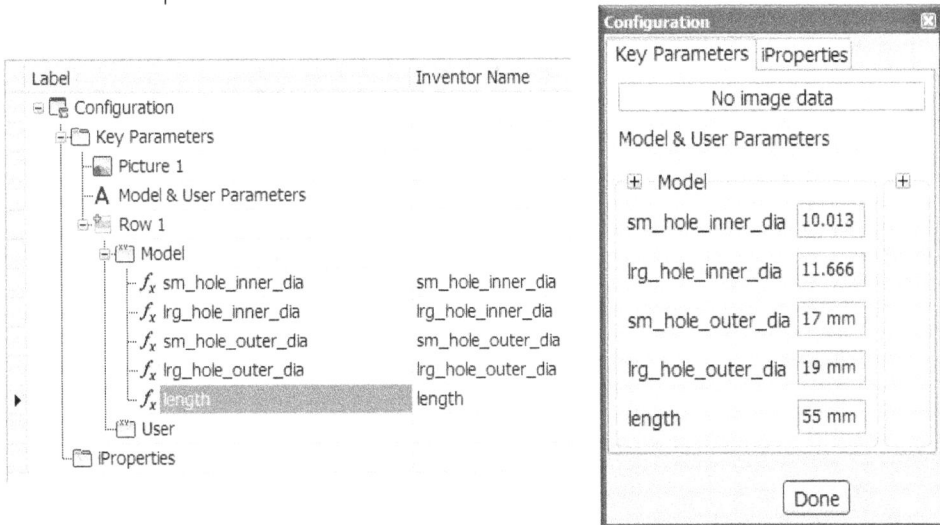

Label	Inventor Name
⊟ 🖳 Configuration	
⊟ 🗀 Key Parameters	
🖼 Picture 1	
A Model & User Parameters	
⊟ 🔳 Row 1	
⊟ (xv) Model	
ƒₓ sm_hole_inner_dia	sm_hole_inner_dia
ƒₓ lrg_hole_inner_dia	lrg_hole_inner_dia
ƒₓ sm_hole_outer_dia	sm_hole_outer_dia
ƒₓ lrg_hole_outer_dia	lrg_hole_outer_dia
ƒₓ length	length
(xv) User	
🗀 iProperties	

Configuration

Key Parameters | iProperties

No image data

Model & User Parameters

⊞ Model ⊞

sm_hole_inner_dia `10.013`

lrg_hole_inner_dia `11.666`

sm_hole_outer_dia `17 mm`

lrg_hole_outer_dia `19 mm`

length `55 mm`

[Done]

Figure 7–16

5. Continue to drag and drop controls from any of the tabs into the Design Tree to complete the form.

- When Rules are added as controls in a form, buttons are added, similar to the **DutyType_Material** rule shown in Figure 7–17. To execute the Rule from a form, select the button.

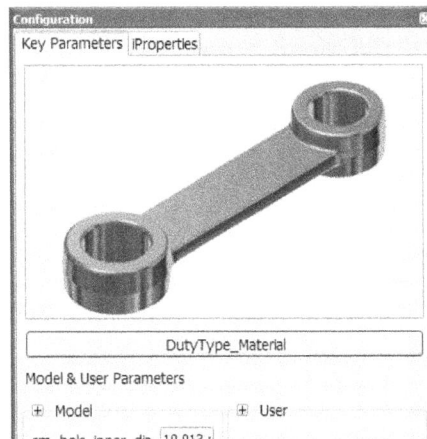

Configuration

Key Parameters | iProperties

DutyType_Material

Model & User Parameters

⊞ Model ⊞ User

sm_hole_inner_dia 10.013 r

Figure 7–17

- When iProperties are added as a control in a form, they are added as entry fields, as shown in Figure 7–18. The *iProperties* tab contains a full list of iProperties that can be dragged into a Form. If an iProperty value is already assigned the value is displayed.

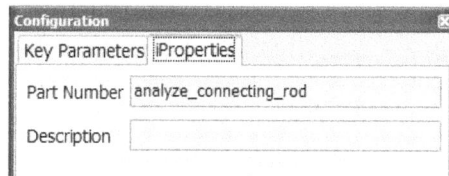

Figure 7–18

Step 4 - Customize the Properties of the form components.

Once controls have been incorporated into the form layout, you can further customize the look of the form by editing properties. Properties exist for both the toolbox components and the controls that were selected in the tabs area.

How To: Customize Properties

1. Select a toolbox component or a control that has been added to the *Design Tree* area. The available properties that pertain to the selected item update in the *Properties* area in the bottom right of the Form Editor dialog box.

   - The properties shown in Figure 7–19 are those that are available for the Tab Group Toolbox component. The field shown at the top of the *Properties* area indicate which item's properties are being displayed.

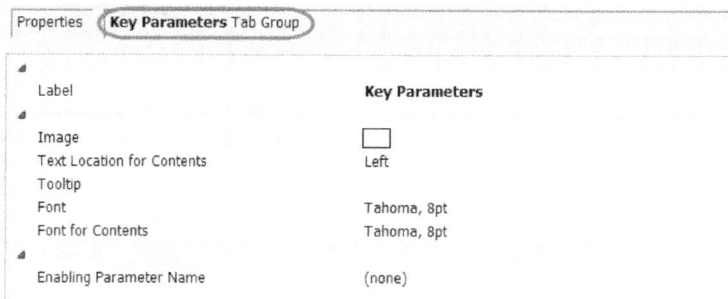

Figure 7–19

2. Select a property value in the right side column to make changes. For example, for the Tab Group, the Font type and size can be selected and changed. Additionally, the text location for its contents can be controlled. The available list of properties varies depending on the selected item in the Design Tree.

- The properties shown in Figure 7–20 are those available for the Picture component. In order to add an image to the form, the image must be added in the *Properties* area. In the Image row, click ⊡ (browse) and select the image file to be added. In this case, the **connecting_rod.png** file was added. The Preview Form is shown in Figure 7–21.

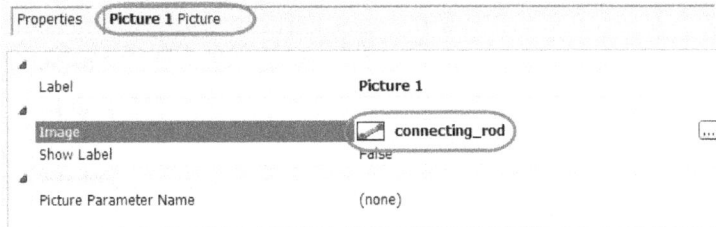

Figure 7–20

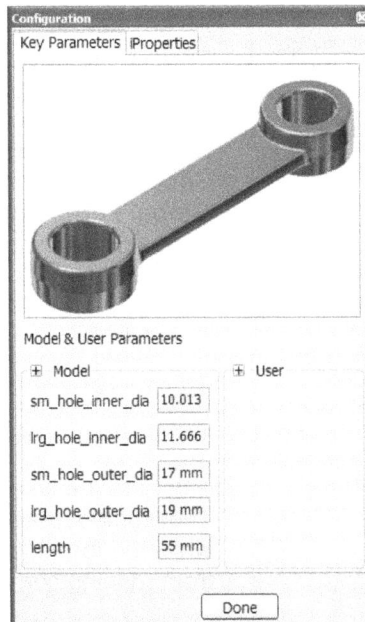

Figure 7–21

Step 5 - Complete form Creation.

To complete the form once the layout and controls have been customized, click **OK**. The Form Editor and the Preview dialog boxes are closed and the form is added to the *Forms* tab in the iLogic browser, as shown in Figure 7–22.

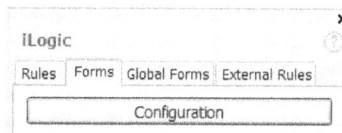

Figure 7–22

Step 6 - Verify the Form.

Once a form has been completed, it is a good practice to test it to ensure that it provides all the required information and works as expected. To open the form, on the *Forms* tab in the iLogic browser, select the form name. The form opens and you can enter values for the listed controls to customize the model.

As an alternative to explicitly executing a form in the iLogic browser, you can also add code to a rule that launches a form. The *Forms* category in the *System* tab in the *Snippets* area provides six snippets that can be used. Three of the snippets are for local forms (**Show Form, Show Form (non-modal)**, and **Show Form (modal)**) and three are for global forms (**Show Global Form, Show Global Form (non-modal)**, and **Show Global Form (modal)**).

```
iLogicForm.Show("Form 1")
iLogicForm.Show("Form 1", FormMode.NonModal)
iLogicForm.Show("Form 1", FormMode.Modal)
iLogicForm.ShowGlobal("Form 1")
iLogicForm.ShowGlobal("Form 1", FormMode.NonModal)
iLogicForm.ShowGlobal("Form 1", FormMode.Modal)
```

The Form1 element represents the local or global form name. The NonModal element enables you to run other commands while the form is displayed. The Modal element prevents you from running commands until the form is closed.

Figure 7–23 shows the Configuration form when the Length value has been modified from 50 mm to 90 mm using the Form's interface. The model updates as expected and the overall length of the connecting rod updates.

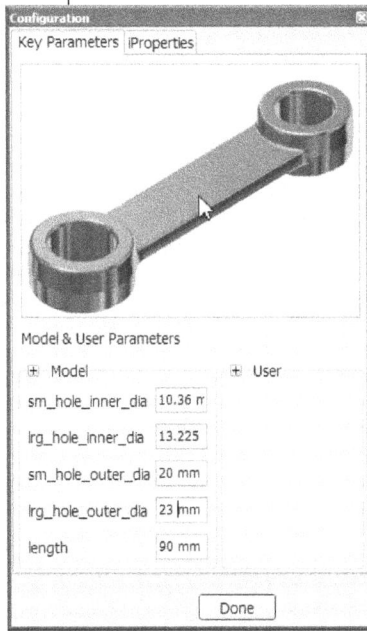

Figure 7–23

Step 7 - Edit the form, as required.

If changes are required, you edit an existing form by changing the layout or the control items that have been added to the layout components.

- If changes are required to the form, right-click the form name in the iLogic browser and select **Edit**, as shown in Figure 7–24.

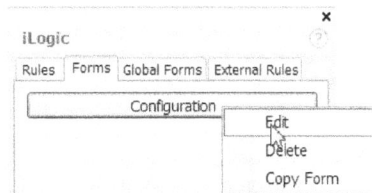

Figure 7–24

Once you have selected the **Edit** option, the Form Editor opens and you can make changes to the form by right-clicking any of the items and selecting **Delete** to remove them or you can drag and drop them in the Design Tree to reposition them to change the layout. Additional toolbox components and controls can also be added.

- To delete a form, right-click the form name in the iLogic browser and select **Delete**.

- To copy an existing form, right-click the form name in the iLogic browser and select **Copy Form**. To paste the copied form, right-click again in the *Forms* tab of the iLogic browser and select **Paste Form**. Copying forms is useful if you require a similar second form. Copying and pasting forms is also supported between documents.

Practice 7a

Creating Triggers and Forms

Practice Objective

- Review a model's existing parameters and rules to gain an understanding of how the model can be manipulated.
- Set an event trigger to automatically launch an iLogic rule when a material change is made in the model.
- Create an iLogic custom form where you can enter parameter and iProperty data for the model.
- Launch an iLogic form using the iLogic browser and based on the use of an iTrigger.
- Enter parameter and iProperty data for a model using a custom iLogic form.

In this practice you will work with an existing model and create triggers that can automatically execute an iLogic Rule. The triggers will be created as an event trigger and an iTrigger. Additionally, you will create a custom form that enables you to enter parameter and iProperty data for the model. The iTrigger that is created will execute a rule that in turn executes the form.

Task 1 - Open and review an existing model.

In this task, you open a model and review its model geometry, parameters, and rules to gain an understanding of how the model can be manipulated.

1. Open the model **Sprocket.ipt** from the top-level practice files folder. The model displays as shown in Figure 7–25.

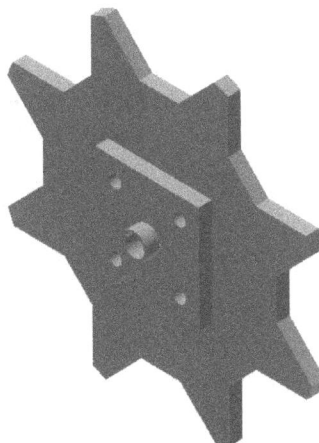

Figure 7–25

2. In the Quick Access Toolbar, click fx (Parameters). The Parameters dialog box opens.

3. Filter the parameter list using the **Renamed** option. The Parameters dialog box updates as shown in Figure 7–26. The model parameters that have been renamed from their default names are displayed. These are the parameters that can be used to drive changes to the size of the model. The User parameter is a multi-value parameter that is used in conjunction with iLogic to drive the material that is assigned in the model. Close the Parameters dialog box.

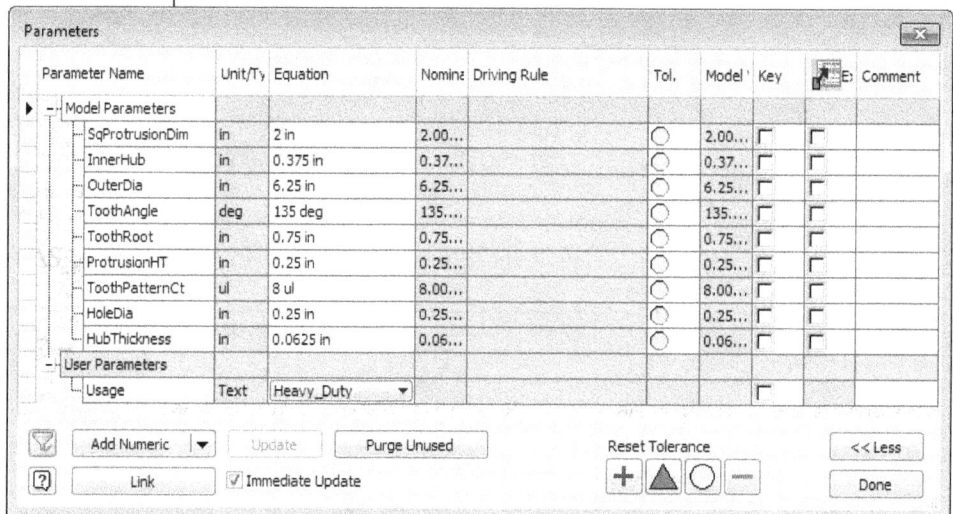

Figure 7–26

4. In the iLogic browser, right-click on the **Material** rule and select **Edit Rule**. The Rule Editor opens and lists the lines of code, as shown in Figure 7–27. This rule assigns the material for the model based on the value that is selected for the **Usage** parameter. Close the Rule Editor.

```
If Usage = "Heavy_Duty" Then
iProperties.Material = "Steel, High Strength Low Alloy"

ElseIf Usage = "Medium_Duty" Then
iProperties.Material = "Steel, Carbon"

ElseIf Usage = "Light_Duty" Then
iProperties.Material = "Steel, Mild"

End If
```

Figure 7–27

5. Open the Parameters dialog box and change the **Usage** parameter value to **Medium_Duty**. If the material type does not update in the Quick Access Toolbar, right-click the filename in the Model browser and select **iProperties**. Select the *Physical* tab and verify that the **Material** is set as **Steel, Carbon**. Close the dialog box. In the Autodesk Inventor 2017 software the material does not always correctly update in the Quick Access Toolbar.

Task 2 - Set an Event Trigger.

In this task, you set an event trigger so that if the material for the model is manually changed, the Material rule will execute. This is important to ensure that the material type is tied to the **Usage** parameter setting so that it cannot be independently changed.

1. In the Quick Access Toolbar, select the Material drop-down list and select the **Steel, Cast** material. This is not one of the material types that is to be used for the Heavy, Medium, and Light Duty usages. However, it is possible for you to reset the material.

Open and close the iProperties dialog box again to ensure that the material updates correctly.

2. To reset the material, right-click on the **Material** rule in the iLogic browser and select **Run Rule**. The correct material should now be assigned.

3. In the *Manage* tab>iLogic panel, click (Event Triggers) to create a trigger.

*Alternatively, you can double click on the Event name or right-click and select **Select Rules** to assign the rules to an event.*

4. Select the **Material Change** event in the Rules Triggered by Events dialog box and click **Select Rules**. The Choose Rules for Material Change dialog box opens.

5. Select the **Material** rule so that it is assigned to the Material Change trigger, as shown in Figure 7–28.

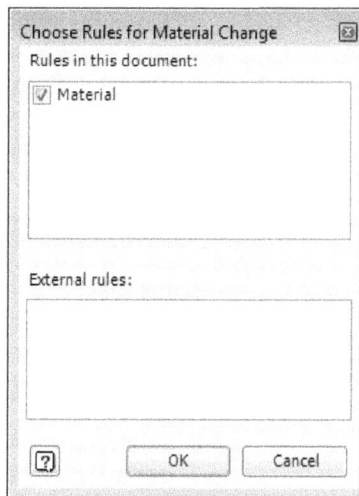

Figure 7–28

6. Click **OK**. Note that the **Material** rule is now listed under the Material Change event trigger, as shown in Figure 7–29.

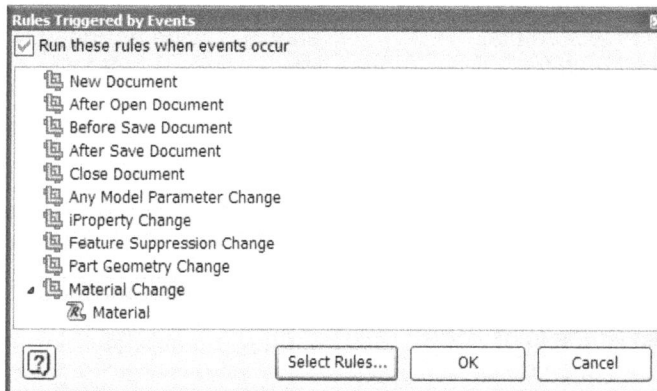

Figure 7–29

7. Click **OK** to close the Rules Triggered by Events dialog box.

8. In the Quick Access Toolbar, select the Material drop-down list and again select the **Steel, Cast** material. Note how you cannot change the material type. As soon as a manual material change is made to the model, the **Material** rule is executed because of the event trigger and it is reset.

Task 3 - Create a Form to enter parameter data.

In this task, you create a form that contains a picture of the model, three tab groups, and a standard group.

1. In the *Manage* tab>expanded iLogic panel, click ⬚ (Add Form).

2. Enter **Sprocket** as the name of the form and maintain the default setting for being created in this document. Click **OK**.The Form Editor and Form preview dialog boxes open.

3. In the *Toolbox* area, select and drag the **Picture** (⬚) component to the Design Tree. The Design Tree updates as shown in Figure 7–30.

Label	Inventor Name
⊟ ⬚ Sprocket	
I └⬚ Picture 1	

Figure 7–30

4. Select the Picture 1 line in the Design Tree. In the *Properties* area of the Form Editor dialog box, select the **Image** property, and click ⬚ (Browse). Navigate to the practice files folder and select and open **Sprocket.png**. The image is displayed in the Form Preview dialog box, as shown in Figure 7–31.

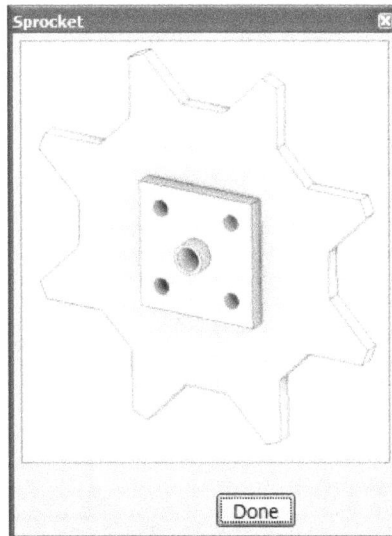

Figure 7–31

5. In the *Toolbox* area, select and drag three **Tab Group** (□)
   components and one **Group** (□) component to the Design
   Tree. The Design Tree updates as shown in Figure 7–32.

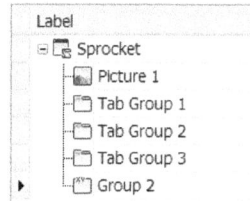

Figure 7–32

*Alternatively, you can
edit the Properties of
each of the components
and edit the headings in
the Label property.*

6. Double-click each of the Tab Group and Group headings and
   rename them as shown in Figure 7–33. The Form Preview
   dialog box updates as shown in Figure 7–34.

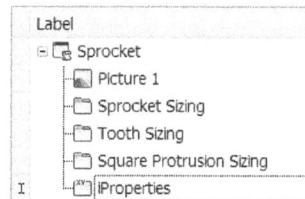

Figure 7–33

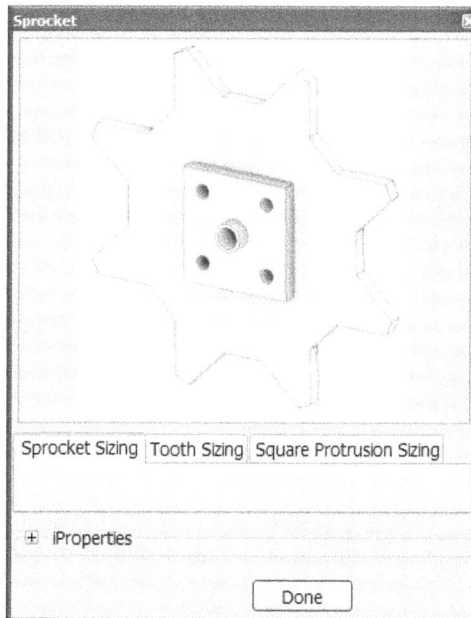

Figure 7–34

7. The filter setting (**Renamed**) that was set in the Parameters dialog box should be maintained in this tab.

8. Select the *Parameters* tab in the *Tabs* area of the Form Editor dialog box.

9. Press and hold <Ctrl> and select the **OuterDia**, **InnerHub**, and **HubThickness** parameters. Drag the three parameters from the *Parameters* tab to the Sprocket Sizing Tab Group, as shown in Figure 7–35.

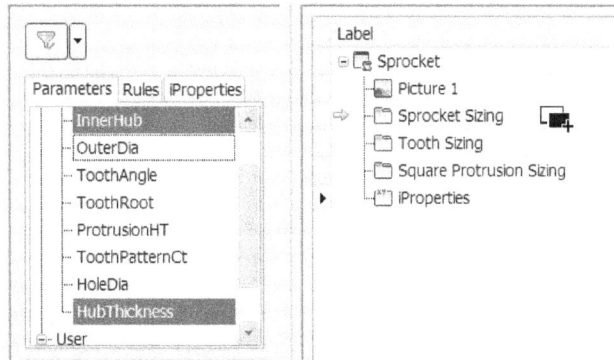

Figure 7–35

10. Release the mouse to place the parameters in the Tab Group. The bottom of the Form Preview dialog box updates as shown in Figure 7–36. Drag to reorder the list, if required.

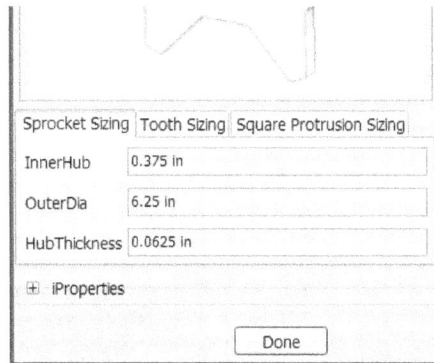

Figure 7–36

11. Often the parameter names that are used in the model might not be descriptive enough. The parameter display name in the form can be edited. Select the **OuterDia** parameter in the Design Tree.

12. In the *Properties* area of the Form Editor dialog box, select the label value and enter **Sprocket OD**, as shown in Figure 7–37. This is the name of the parameter in the form but it remains tied to the **OuterDia** parameter in the model.

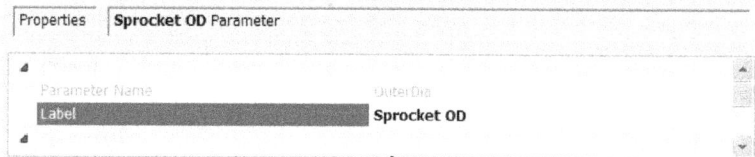

Figure 7–37

13. As an alternative to changing the label for a parameter in the *Properties* area, select the parameter name in the left column of the Design Tree (not the *Inventor Name* column) and edit the parameter name here. Edit the **InnerHub** parameter to **Inner Hub OD**, and the **HubThickness** parameter to **Hub Thickness**. The renamed parameter labels are shown in Figure 7–38.

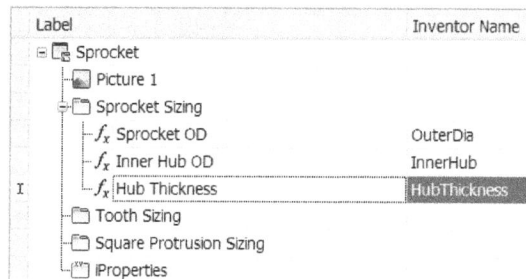

Figure 7–38

To clear previously selected parameters, select a single parameter and then use <Ctrl> to select additional parameters.

14. In the *Parameters* tab, hold <Ctrl> and select the **ToothAngle**, **ToothRoot**, and **ToothPatternCT** parameters. Drag the three parameters from the *Parameters* tab to the Tooth Sizing Tab Group.

15. Edit the parameter labels for each of the **Tooth** parameters, as shown in Figure 7–39. Use the *Properties* area or make the change in the Design Tree.

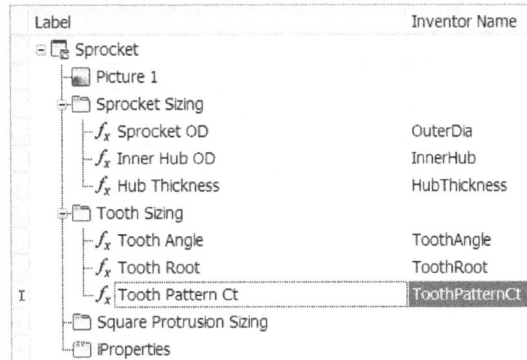

Label	Inventor Name
⊟ ⬚ Sprocket	
🖻 Picture 1	
⊟ 🗀 Sprocket Sizing	
f_x Sprocket OD	OuterDia
f_x Inner Hub OD	InnerHub
f_x Hub Thickness	HubThickness
⊟ 🗀 Tooth Sizing	
f_x Tooth Angle	ToothAngle
f_x Tooth Root	ToothRoot
I f_x Tooth Pattern Ct	ToothPatternCt
🗀 Square Protrusion Sizing	
📰 iProperties	

Figure 7–39

16. In the *Parameters* tab, press and hold <Ctrl> and select the **SqProtrusionDim**, **ProtrusionHT**, and **HoleDia** parameters. Drag the three parameters from the *Parameters* tab to the Square Protrusion Sizing Tab Group.

17. Edit the parameter labels for each of the parameters, as shown in Figure 7–40. Use the *Properties* area or make the change in the Design Tree.

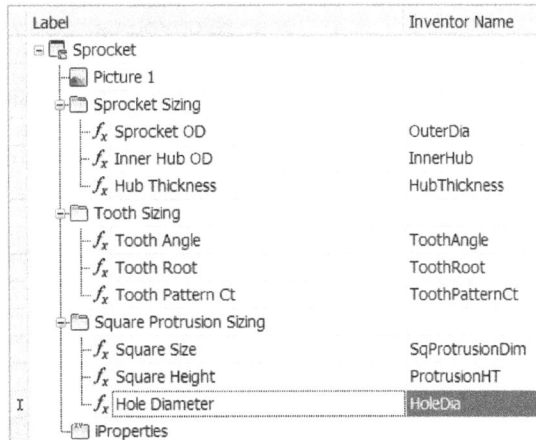

Label	Inventor Name
⊟ ⬚ Sprocket	
🖻 Picture 1	
⊟ 🗀 Sprocket Sizing	
f_x Sprocket OD	OuterDia
f_x Inner Hub OD	InnerHub
f_x Hub Thickness	HubThickness
⊟ 🗀 Tooth Sizing	
f_x Tooth Angle	ToothAngle
f_x Tooth Root	ToothRoot
f_x Tooth Pattern Ct	ToothPatternCt
⊟ 🗀 Square Protrusion Sizing	
f_x Square Size	SqProtrusionDim
f_x Square Height	ProtrusionHT
I f_x Hole Diameter	HoleDia
📰 iProperties	

Figure 7–40

18. Select the *iProperties* tab. All iProperties are listed and are sub-divided based on the tab names in the iProperties dialog box.

19. Press and hold <Ctrl> and select the **Part Number**, **Stock Number**, and **Description** parameters. Drag the three parameters from the *Parameters* tab to the iProperties Group, as shown in Figure 7–41.

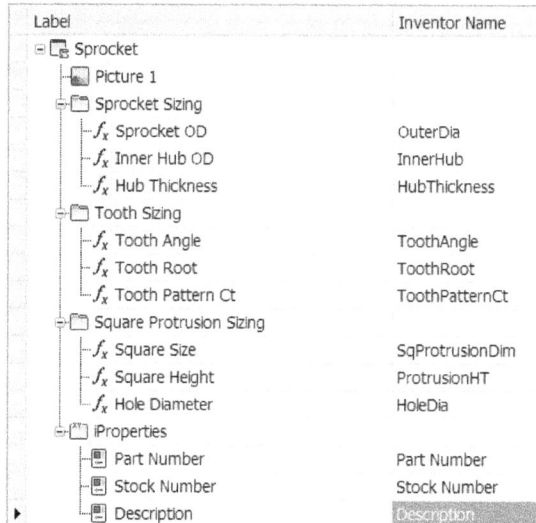

Label	Inventor Name
⊟ Sprocket	
— Picture 1	
⊟ Sprocket Sizing	
— f_x Sprocket OD	OuterDia
— f_x Inner Hub OD	InnerHub
— f_x Hub Thickness	HubThickness
⊟ Tooth Sizing	
— f_x Tooth Angle	ToothAngle
— f_x Tooth Root	ToothRoot
— f_x Tooth Pattern Ct	ToothPatternCt
⊟ Square Protrusion Sizing	
— f_x Square Size	SqProtrusionDim
— f_x Square Height	ProtrusionHT
— f_x Hole Diameter	HoleDia
⊟ iProperties	
— Part Number	Part Number
— Stock Number	Stock Number
— Description	Description

Figure 7–41

Currently the form only has a **Done** button on it. The Form dialog box is required to have **Cancel** and **Apply** buttons as well, in case a change is made in error and you can cancel out of the change or you can apply changes and continue to enter values.

20. At the top of the Design Tree, select **Sprocket** (⬚). Scroll down in the *Properties* area of the Form Editor dialog box and select the **Predefined Buttons** property. In the drop-down list, select **OK Cancel Apply**, as shown in Figure 7–42.

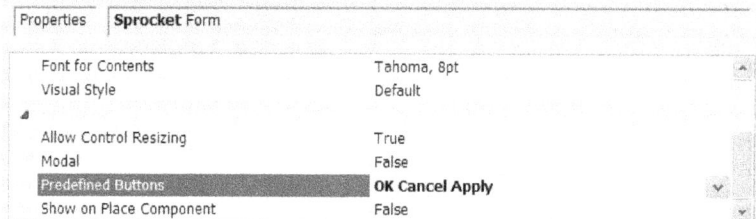

Properties	**Sprocket** Form
Font for Contents	Tahoma, 8pt
Visual Style	Default
Allow Control Resizing	True
Modal	False
Predefined Buttons	**OK Cancel Apply**
Show on Place Component	False

Figure 7–42

The Form Preview dialog box updates as shown in Figure 7–43.

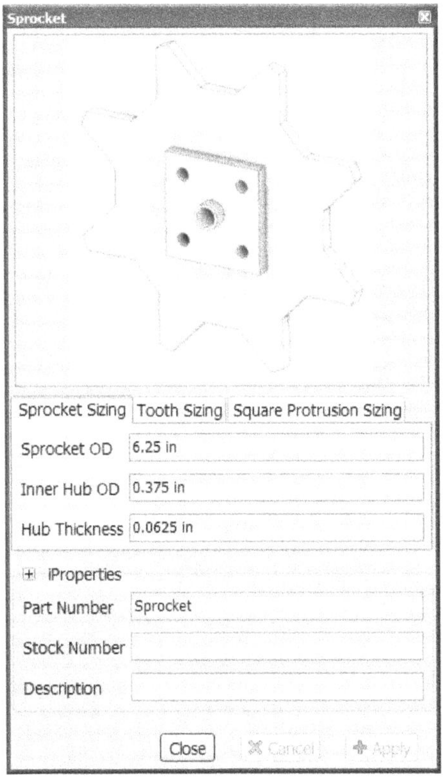

Figure 7–43

21. To complete the form, click **OK**. The Form Editor and the Preview dialog boxes are closed and the form is added to the *Forms* tab in the iLogic browser, as shown in Figure 7–44.

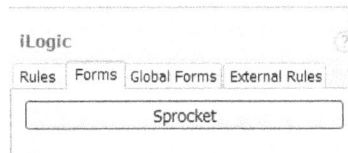

Figure 7–44

22. The **Usage** parameter has not been included in the form. Right-click **Sprocket** in the *Forms* tab of the iLogic browser and select **Edit**.

23. Select the **Usage** parameter in the *Parameters* tab. Drag it directly beneath the Picture component near the top of the Design Tree. If the Usage control is dropped in another location, reselect it and drag it into position, as shown in Figure 7–45. The bottom portion of the form updates, as shown in Figure 7–46.

Repositioning in the list enable for more specific placement than when you initially place it in the Design Tree.

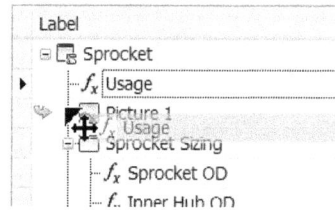

Figure 7–45

Figure 7–46

24. Click **OK** to complete the form.

Rules can also be added to a form; however, they are not required in this model.

Task 4 - Run the Form.

In this task, you open the form and makes changes to the displayed parameters.

1. In the iLogic browser, click **Sprocket** to launch the Sprocket Form. The form displays as shown in Figure 7–47.

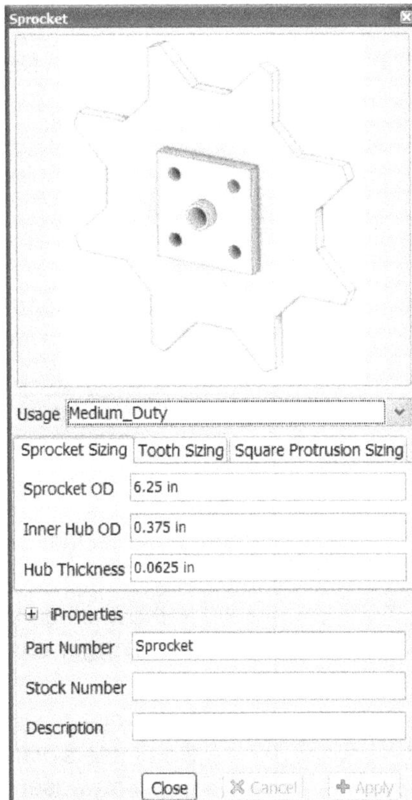

Figure 7–47

2. Using the following table, modify the listed parameter values by making the changes in the Sprocket form. When you make a dimensional change in the model, press <Enter> and **Apply** becomes available. Click **Apply** to update the model while remaining in the form to make additional changes.

Component	Parameter Name	New Value
N/A	Usage	Heavy_Duty
Sprocket Sizing Tab	Inner Hub OD	0.5 in
Sprocket Sizing Tab	Hub Thickness	0.1 in
Tooth Sizing Tab	Tooth Angle	120 deg
Square Protrusion Sizing Tab	Square Size	2.25 in
Square Protrusion Sizing Tab	Hole Diameter	.3 in
iProperties Group	Part Number	Sprocket - Heavy
iProperties Group	Stock Number	63288-SH
iProperties Group	Description	Suitable for Heavy Duty Use

3. Click **OK** to apply the change and return to the newly configured model.

Task 5 - Create an iTrigger to launch the Sprocket Form.

The iLogic browser is not always visible for launching iLogic forms. In this task, you learn how to add an **iTrigger0** parameter to the model that can be used in an iLogic rule to launch a form.

1. In the *Manage* tab>iLogic panel, click ◈ (iTrigger) to create the **iTrigger** parameter in the current document.

2. In the Quick Access Toolbar, click f_x (Parameters). The Parameters dialog box opens, as shown in Figure 7–48. Note that a new user parameter has been added to the model. This is the **iTrigger0** parameter that was added to the model in the previous step.

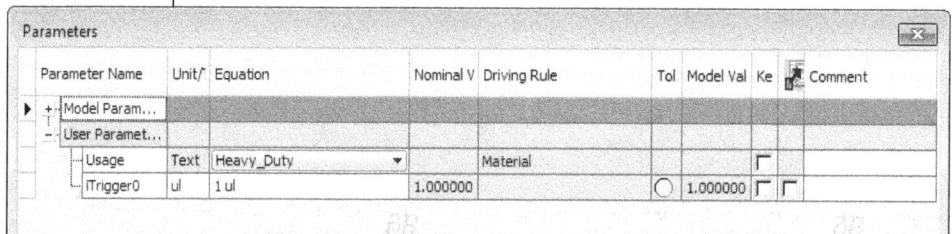

Parameter Name	Unit/	Equation	Nominal V	Driving Rule	Tol	Model Val	Ke	Comment
+ Model Param...								
− User Paramet...								
Usage	Text	Heavy_Duty ▼		Material			☐	
iTrigger0	ul	1 ul	1.000000		◯	1.000000	☐ ☐	

Figure 7–48

3. Close the Parameters dialog box.

4. In the *Manage* tab>iLogic panel, click ⬚ (Add Rule).

5. Enter **FormTrigger** as the name of the rule. The Edit Rule dialog box opens.

6. In the *Rule Editor* area, enter the code shown in Figure 7–49. This code identifies which rules will be executed when the iTrigger is activated.

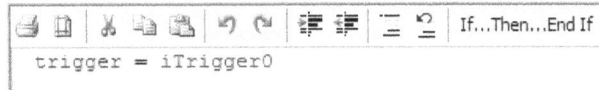

```
trigger = iTrigger0
```

Figure 7–49

7. Currently there are no actions specified in the FormTrigger rule. Add the line of code shown in Figure 7–50. This line of code specifies that when this rule is activated, the iLogic Form called *Sprocket* will be launched.

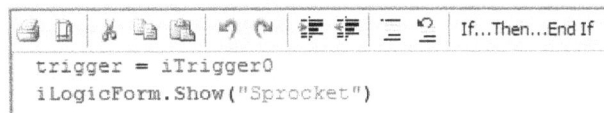

```
trigger = iTrigger0
iLogicForm.Show("Sprocket")
```

Figure 7–50

8. Click **OK** to close the Rule Editor. Close the Sprocket form. The reason it displays is because when you close the Edit Rule dialog box the rule is executed and in this case execution of this rule launches the Sprocket form.

9. In the *Manage* tab>iLogic panel, click ✴ (iTrigger) to execute the iTrigger in the current document. The Sprocket form launches. Launching the form through an iTrigger provides an alternative to always leaving the iLogic browser visible.

10. Close the form without making any additional changes.

11. Save the model and close the window.

Additional rules can be included in the model to provide specific model configurations of the sprocket based on parameter values. If time permits you can add rules to the model to further control sizes for each of the Usage types.

Chapter Review Questions

1. Which of the following are valid Part model event triggers? (Select all that apply.)

   a. New Document

   b. Drawing View Change

   c. Any Model Parameter Change

   d. Material Change

   e. Component Suppression Change

   f. Feature Suppression Change

2. Custom Event Triggers can be created in the Rules Triggered by Events dialog box.

   a. True

   b. False

3. When assigning rules to an Event Trigger, which rules in the document are listed and can be selected from?

   a. Only external rules.

   b. External rules and those document rules that contain the **iTrigger0** parameter line of code.

   c. External rules and those rules that contain lines of code that reference the parameters in the model.

   d. External rules and all rules in the current document.

4. Which of the following is the parameter that is created when ✧ (iTrigger) is executed for the first time in a model?

   a. **iTrigger**

   b. **iTrigger0**

   c. **iTrigger1**

   d. **iTrigger2**

5. Which of the following is the correct syntax for the single line of code that is required in a rule so that the same rule is executed when ✧ (iTrigger) is executed?

   a. iTrigger0

   b. iTtrigger0 = trigger

   c. trigger = iTrigger0

   d. iTtrigger0 = iTtrigger1

6. iLogic forms can only be added to a model that contains iLogic rules.

   a. True

   b. False

7. When creating a form using the Form Editor, which area in the dialog box (shown in Figure 7–51) enables you add the image file that is to populate the *Picture* field in the form?

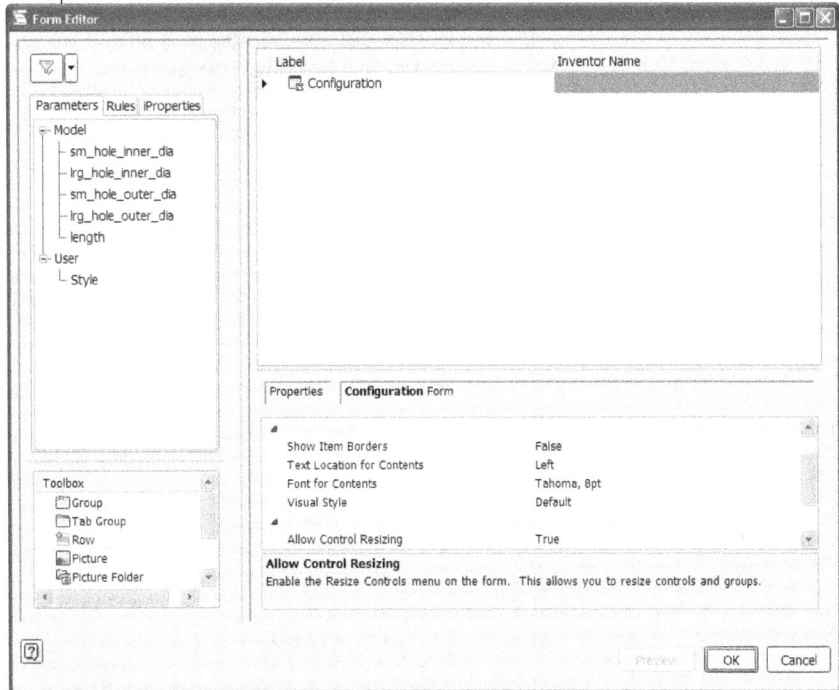

Figure 7–51

   a. *Parameters* tab

   b. *Tool Box* (Picture component) area

   c. *Design Tree* area

   d. *Properties* area

8. Match the Toolbox component on the right with their component type.

a. Group	1.	
b. Tab Group	2.	A
c. Row	3.	
d. Picture	4.	
e. Picture Folder	5.	
f. Empty Space	6.	
g. Label	7.	
h. Splitter	8.	

9. Forms can be copied in a single document, but they cannot be copied between documents.

   a. True

   b. False

10. Which of the following Design Tree configurations is used to create the form in Figure 7–52?

Figure 7–52

a.

b.

c.

d.

Command Summary

Button	Command	Location
	Add Rule	• **Ribbon:** *Manage* tab>iLogic panel • *(right-click in the Rules tab in the iLogic Browser)*
	Event Trigger	• **Ribbon:** *Manage* tab>iLogic panel
	Form	• **Ribbon:** *Manage* tab>expanded iLogic panel
	iTrigger	• **Ribbon:** *Manage* tab>iLogic panel

Miscellaneous Features & Functions

There are additional rule options, search and replace functionality, and the wizards available in the Edit Rule dialog box that can be used to help you be more efficient in creating and editing code for a rule. Additionally, there are some miscellaneous functions that can be used in your models to help automate a design.

Learning Objectives in this Chapter

- Specify a rule behavior to control rule suppression, execution when changes are made, and dialog box display when a rule is created or edited.
- Find and replace a defined string in both the current rule and all rules in the document.
- Create code that generates a Message box when its rule is executed.
- Create code that changes the model's orientation and zoom level to that of when the code was created.
- Create code that prompts you when a parameter value is outside a defined range.
- Set and read the sheet metal rule and the flat pattern extents in the X-, Y-, and Z-directions using an iLogic function.
- Customize a message box to display information during rule execution.
- Measure values in a file using an iLogic function.

8.1 Rule Options

A rule's behavior can be customized using options in the *Options* tab of the Edit Rule dialog box, as shown in Figure 8–1.

Figure 8–1

The Behavior and Rule Editor options are as follows:

Option	Description
Suppressed	Enables you to define the rule as suppressed. Suppressed rules are not run automatically; however, it can be explicitly run by right-clicking and selecting **Run Rule**. Setting this option is an alternative to explicitly suppressing the rule once it is created.
Silent operation	Enables the rule to run without displaying any dialog boxes that might display. This is done by accepting the default options in the dialog boxes.
Fire dependent rules immediately	Enables you to immediately execute a second rule if a parameter in the current running rule is changed and it is referenced in the second rule. This option enables you to avoid waiting to run the second rule.
Don't run automatically	Enables you to prevent a rule from running when a parameter value is referenced in the rule and the value is changed. To run a rule that has been defined with this option, you must explicitly run the rule.
Straight VB Code	Enables you to identify the rule as code that can be shared between rules.
Select Font	Enables you to customize the font used in the Rule Editor area of the dialog box.
Use Component Names (Assembly rules only)	Sets how **Capture Current State** returns information. • If enabled, the parameter statement is included with the component and parameter name (e.g., Parameter("piston:1", "dia") = 10). • If disabled, the parameter statement is not included (e.g., piston.ipt.dia = 10).

8.2 Search and Replace

Rules can become very large. If a change needs to be made, the Search and Replace functionality can be used instead of manually reading the code and making changes. The Search and Replace functionality is located in the *Search and Replace* tab of the Edit Rule dialog box, as shown in Figure 8–2.

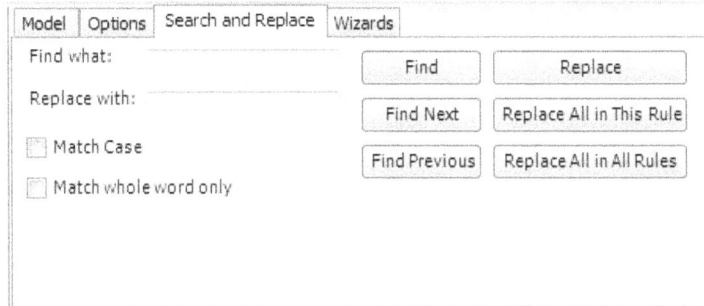

Figure 8–2

Search and Replace works on commented and uncommented code in a rule.

The options for searching and replacing a string in a rule are similar to those found in standard find and replace functionality. You begin by defining the string that you are looking for along with its replacement string. Before searching, you can refine what is found by selecting the **Match Case** or **Match whole word only** options.

Once you begin the search, you can:

- Individually find each instance and replace it.

- Replace all instances in the current rule.

- Replace all instances in all rules in the current document. If you use the **Replace All in All Rules** option, you are prompted with how many rules other than the current were changed.

- Navigate to the next instance (**Find Next**) or return to the previous instance (**Find Previous**).

8.3 Wizards

iLogic provides a predefined interface that enables you to quickly create several different types of code. The code can be used to create a new rule or be used as part of an existing rule. This is accomplished through the use of wizards. These wizards provide an easy-to-use interface that requires user entry to define the required lines of code for a rule. The wizards are included in the *Wizards* tab, as shown in Figure 8–3.

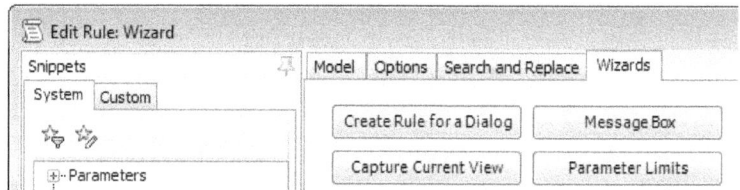

Figure 8–3

The three wizards discussed are described as follows:

Wizard	Description
Capture Current View	Creates a rule that captures the current view state (orientation and zoom settings).
Message Box	Creates a rule that displays a custom message box.
Parameter Limits	Creates a rule that sets the minimum and maximum values for parameters.

The **Create Rule for a Dialog** wizard is not covered in this student guide. Dialog box creation is more commonly done using a custom form. This wizard is meant for more advanced use of forms that are beyond the current functionality of the iLogic Form Editor. In this case, a product like MS VB Express can be used to design a form and compile the code into a DLL. The Create Rule for a Dialog wizard helps connect the DLL and the Autodesk Inventor software using advanced iLogic language.

General Steps

Use the following general steps to use a predefined wizard to create an iLogic rule:

1. Initiate code creation using a wizard.
2. Define the options in the Wizard dialog box.
3. Verify the code.

Step 1 - Initiate code creation using a wizard.

A wizard enables you to add predefined code in a rule. To add the code, you can create a new rule or you can edit an existing rule. In the Edit Rule dialog box, select the *Wizards* tab and select the required wizard button, as shown in Figure 8–4.

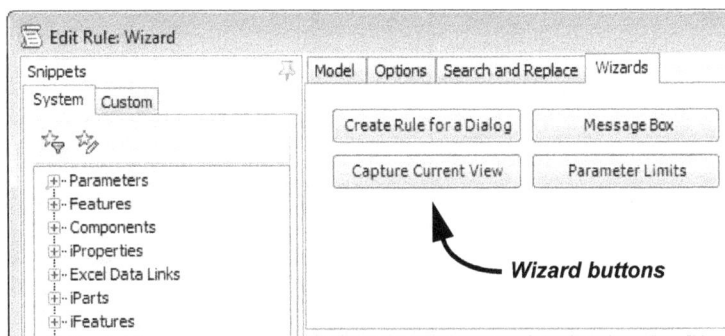

Figure 8–4

- To add the code in the required location in an existing rule, place the cursor in the required location and then select the wizard button.

- If it is not located correctly, it can be copy and pasted after it has been created.

Step 2 - Define the options in the Wizard dialog box.

Once a wizard is activated, its dialog box opens where you can define the required options. The options available depends on the type of wizard that was selected.

Message Box

Clicking **Message Box** opens the MessageBox Wizard dialog box, as shown in Figure 8–5.

Figure 8–5

- Enter a name for the message box in the *Title* field.

- Enter the text that you want to be displayed as a prompt when the rule is executed and the message box is displayed.

- Select the button layout in the Buttons drop-down list, as shown in Figure 8–6. This defines which buttons display at the bottom of the Message Box dialog box when it is displayed.

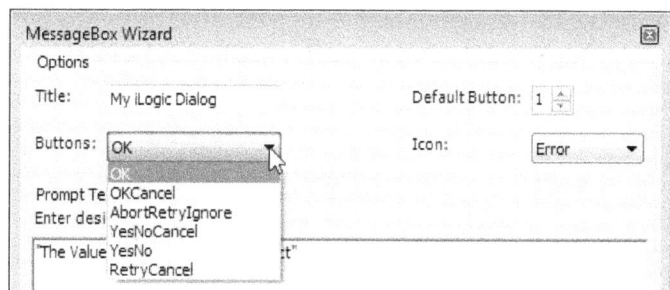

Figure 8–6

- Use the *Default Button* value list to define which button is the default button in the dialog box. The *Default Button* value corresponds to the command order in the list. For example, with the **Buttons** option set as **YesNoCancel**, if you select **2** as the *Default Button* value, then **No** is set as the default button.

- Use the Icon drop-down list (shown in Figure 8–7) to assign the status icon that displays in the dialog box. For example, use the **Information** icon to display ⓘ before the prompted text.

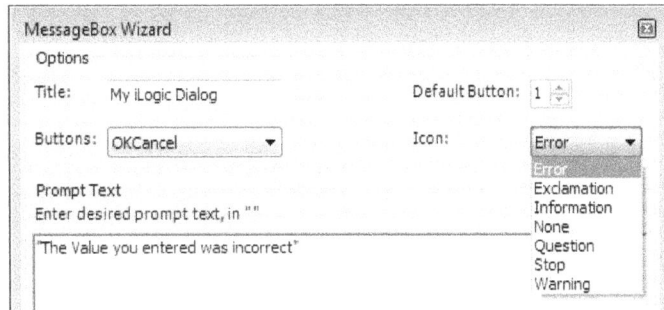

Figure 8–7

To complete the rule, click **OK** to close the MessageBox Wizard dialog box. The custom code is displayed in the rule.

Capture Current View

Clicking **Capture Current View** opens the Capture View dialog box, as shown in Figure 8–8.

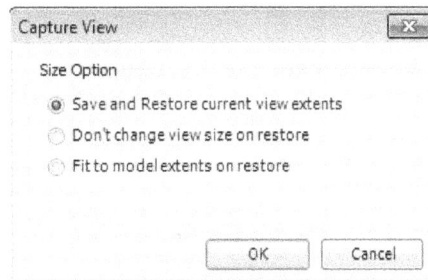

Figure 8–8

To complete the rule that captures the current view of the model, select one of the three options:

- The **Save and Restore current view extents** option creates a rule that can be executed to restore the view to the model's orientation and zoom settings that existed when the rule was captured.

- The **Don't change view size on restore** option creates a rule that can be executed to move the model to the center of the display, resets the orientation to that when the rule was created, but does not change the zoom level.

- The **Fit to model extents on restore** option creates a rule that can be executed to move the model to the middle of the display and at the same time fills the display (similar to **Zoom All**). It does not change the orientation of the model.

To complete the rule, click **OK** to close the Capture View dialog box. The custom code displays in the rule.

Parameter Limits

Clicking **Parameter Limits** opens the iLogic Limits Wizard dialog box, as shown in Figure 8–9. This wizard enables you to create code that sets acceptable limits for a parameter value.

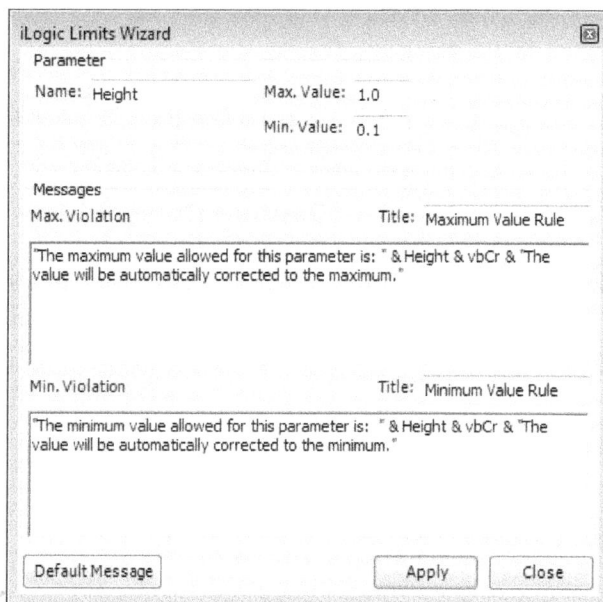

Figure 8–9

Refer to the Parameters dialog box for the correct syntax for the parameter.

- Enter the parameter name that is to be controlled by limits.

- Enter a min and max value for the parameter.

- Enter custom messages that will be displayed if a value is not within the assigned range. The name of the dialog box can be provided in the *Title* field and a custom message in the *Max/Min Violation* areas.

To complete the rule, click **Apply** in the iLogic Limits Wizard dialog box and then click **Close**.

Step 3 - Verify the code.

Any rule that was created using a wizard, is listed in the iLogic browser, just like any other rule. The specific instructions to execute each wizard type are described below.

Message Box

To verify the code that is generated by the Message Box wizard, you can explicitly run the rule in which it resides. Alternatively, make changes to the model that causes the execution of the rule. The Message Box (shown in Figure 8–10) is an example of a custom dialog box that can be created.

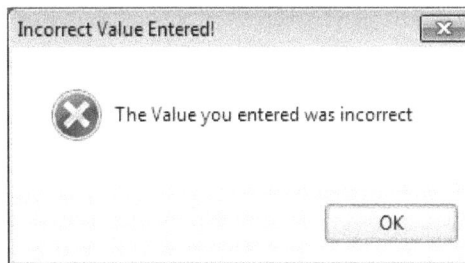

Figure 8–10

Capture Current View

To verify the code that is generated by the Capture Current View wizard, reorient and zoom the model to a new orientation and execute the rule in which the code resides to reorient, as required.

- A rule that was created with the **Save and Restore current view extents** option in the Capture Current View Wizard reorients the model back to the model's orientation and zoom settings that existed when the rule was captured.

- A rule that was created with the **Don't change view size on restore** option in the Capture Current View Wizard moves the model to the center of the display and resets the orientation to that when the rule was created.

- A rule that was created with the **Fit to model extents on restore** option in the Capture Current View Wizard moves the model to the middle of the display and fills the display (similar to **Zoom All**).

Parameter Limits

To verify the code that is generated by the Parameter Limits wizard, open the Parameters dialog box and enter a value that is outside the range of that defined in the code. A message displays in the form of a dialog box, similar to that shown in Figure 8–11.

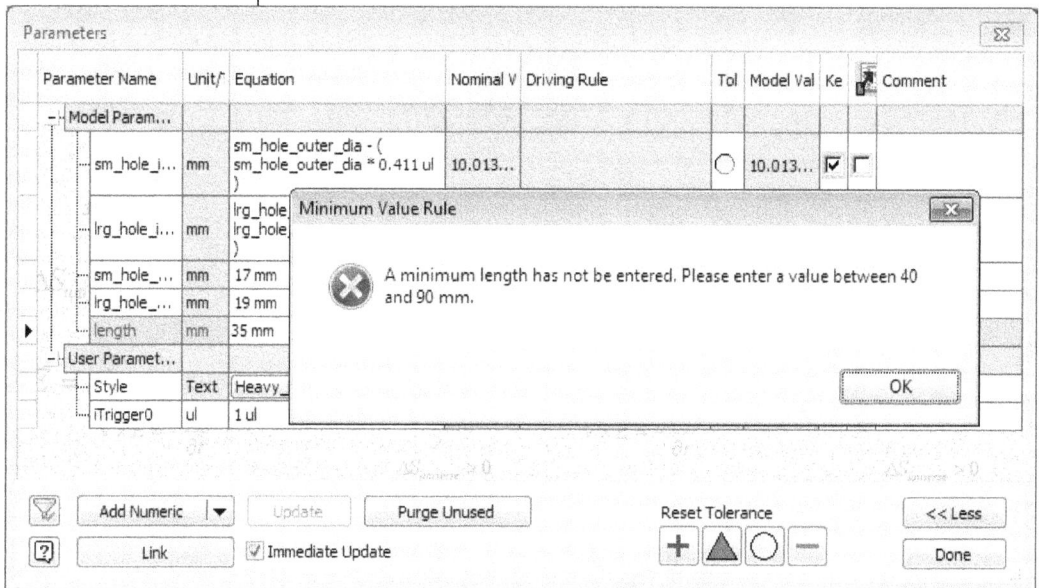

Figure 8–11

8.4 Miscellaneous Functions

In this student guide, the functions most commonly used in part, assembly, and drawing rules have been discussed. The following describes additional functions that can be used.

Sheet Metal

The *Sheet Metal* category of functions enables you to read and set the active style, read the Kfactor value that is used in the model, and read the X, Y, and Z values of the flat pattern.

- The sheet metal rule that is assigned in a model specifies many default settings for model design. Using the **Set Active Style** function, you can assign the rule to be used in the model. The syntax for this function is as follows.

      SheetMetal.SetActiveStyle("styleName")

- The **Get Active Style** and **Get Active KFactors** functions are read functions that assign the current rule and KFactor value to a variable for use in the rule. The syntax for these functions are as follows.

      currentStyle = SheetMetal.GetActiveStyle()
      kFactor = SheetMetal.ActiveKFactor

- The **FlatExtentsLength**, **FlatExtentsWidth**, and **FlatExtentsArea** functions can be used to read the model extents in the X (red axis) and Y (green axis), directions and its area when it is a flat pattern. The following lines of code show the syntax and the variable name to which the value is assigned.

      extents_length = SheetMetal.FlatExtentsLength
      extents_width = SheetMetal.FlatExtentsWidth
      extents_area = SheetMetal.FlatExtentsAreaMessage Box

The *Message Box* category of functions incorporate visual feedback when a rule is running in the form of a dialog box that is displayed. The commonly used snippets enable you to create the message box, customize the buttons and icons that are displayed, and define input fields.

*The **MsgBox** function, that is a VB.NET syntax, can also be used in an iLogic rule. Incorporating this in the code is commonly used as a troubleshooting technique to help display the value of a parameter.*

- The **Show** function provides the foundation of the message box creation. The syntax for this function is as follows:

      MessageBox.Show("Message", "Title")

- The Message portion of the code defines the text that is to be displayed in the message box. The Title portion of the code defines the title bar heading of the message box.

- The button functions (**OK**, **OKCancel**, **RetryCancel**, **YesNo**, **YesNoCancel**, and **AboutRetryIgnore**) enable you to customize the buttons that are displayed at the bottom of the message box. To incorporate this code into the **Show** function, add an additional element in the brackets, as is done for the **YesNo** buttons function shown in the code below. The resulting message box is shown in Figure 8–12.

  MessageBox.Show("Message", "Title", MessageBoxButtons.YesNo)

The Message Box wizard can also be used to create a message box.

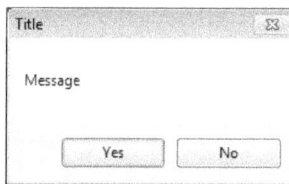

Figure 8–12

- The icon functions (**Error, Exclamation, Information**, **None**, **Question**, **Stop**, **Warning**) enable you to add images in the message box. You incorporate the code as an element in the brackets, as is done for the Information function shown in the code below. The resulting message box is shown in Figure 8–13.

  MessageBox.Show("Message", "Title", MessageBoxButtons.YesNo, MessageBoxIcon.Information)

Figure 8–13

- To incorporate a parameter or variable value in a message box, you must use the **&** symbol to identify the text as a parameter or variable. In the following examples, review how the code is written and how the information is displayed in the corresponding message boxes, as shown in Figure 8–14 and Figure 8–15.

  ```
  MessageBox.Show("The value of the Length parameter is " & length, "Title")
  ```

  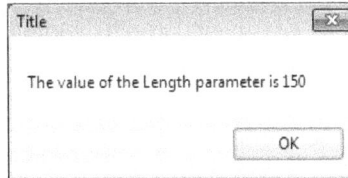

  Figure 8–14

  ```
  MessageBox.Show("Length = " & length & " and Diameter = " & dia, "Title")
  ```

  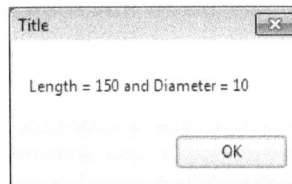

  Figure 8–15

- The **InputBox** function creates an input dialog box that requires user entry instead of an informational message box. The syntax for the **InputBox** function is as follows. The resulting input box is shown in Figure 8–16.

  ```
  myparam = InputBox("Prompt", "Title", "Default Entry")
  ```

*The syntax for the **InputListBox** and **InputRadioBox** functions are similar to the **InputBox** function. They provide alternative input box styles.*

  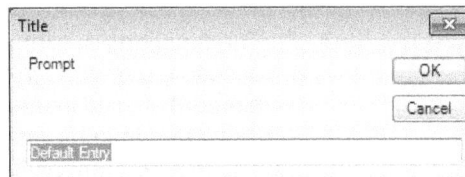

  Figure 8–16

- The **Prompt** parameter enables you to define what is being asked. The **Title** parameter is the title of the input dialog box and the **Default Entry** parameter is the default value that displays when prompted. The value that is entered is assigned to the myparam variable for use later in the rule.

Measure

The *Measure* category of iLogic functions enables you to read measurements in the model. Most of the measurements that are contained in this list involve measuring between work features or iMates; however, you can also measure the area and perimeter of a sketch, and the extents of a model.

- The two **MinimumDistance** and four **Angle** functions included in the list provide measurement options between work features or iMates. The syntax of all of these functions is similar. They require entity names to measure between (entityName#) and in case of an assembly model, you also include the component name to which the measuring reference resides (componentName#). The following two lines of code show the **MinimumDistance** and **Angle (within components)** syntax.

```
Measure.MinimumDistance("entityName1", "entityName2")
Measure.Angle("componentName1", "entityName1",
"componentName2", "entityName2")
```

- The **Area of a Sketch** and **Perimeter** functions conduct measurements on a defined sketch. The following two lines of code show their syntax.

```
Measure.Area("sketchName")
Measure.Perimeter("sketchName")
```

*The **Extents** functions measure between visible entities such as work features, surfaces, or bodies. It cannot measure to hidden entities.*

- The **ExtentsLength**, **ExtentsWidth**, and **ExtentsHeight** functions can be used to read the model extents in the X (red axis), Y (green axis), and Z (blue axis) directions. The following lines of code show their syntax.

```
Measure.ExtentsLength
Measure.ExtentsWidth
Measure.ExtentsHeight
```

Document

The functions available in the *Document* category enable you to access information in the current document. The "ThisDoc" element is used in these functions to indicate that the information requested is to be read for the document that the rule resides in. The most commonly used functions in this category get the filename, update the document, and save it.

If a document is prevented from being updated based on the rule's code, becomes active in the Quick Access Toolbar and must be manually selected to update the model.

- The **UpdateWhenDone** function enables you to force or prevent a document update at the end of rule execution. To force the update, set the function equal to **True**; to prevent the update, use a value of **False**. The syntax for this function is as follows.

      iLogicVb.UpdateWhenDone = True

- The **ThisDoc.Save** function enables you to force a document save after the rule is executed. The syntax for this function is as follows.

      ThisDoc.Save

Run Other Rule

Any Autodesk Inventor model that uses iLogic rules to automate a design might incorporate many different rules to accomplish the design intent. The *Run Other* category of functions enable you to incorporate code in one rule that executes another rule. Incorporating these functions are useful in rules that are not executed with an event trigger, parameter change, or if the **Don't run automatically** option was set when it was created. The syntax for the **RunRule** function is as follows.

      iLogicVb.RunRule("RuleName")

Additionally, it can be used in an assembly to execute rules at a component level. The syntax is similar to executing rules in the same document; however, you must specify the component name.

      iLogicVb.RunRule("PartA:1", "RuleName")

*The **RunRule** function can be added at any point in a rule.*

Chapter Review Questions

1. Suppressed rules cannot be run unless they are resumed.

   a. True

   b. False

2. Which of the following rule options should be used if you want to control the automatic execution of a second rule in which a parameter value was changed based on another rule?

   a. **Fire dependent rules immediately**

   b. **Don't run automatically**

   c. **Silent Operation**

   d. **Use Component Names**

3. The **Search and Replace** functionality can only search for a string in the current rule.

   a. True

   b. False

4. When adding a message box code in a rule, using the Message Box Wizard, how do you set **No** as the default value when the **YesNo** option defines the buttons to be displayed?

   a. Set the **Default Button** option to **Yes**.

   b. Set the **Default Button** option to **No**.

   c. Set the **Default Button** option to **1**.

   d. Set the **Default Button** option to **2**.

5. Which of the following can be customized using the Message Box Wizard? (Select all that apply.)

   a. Title of the Message Box

   b. Buttons displayed

   c. Display time

   d. Icon style

   e. Message

   f. Condition when the Message Box is displayed

6. Which of the following options is used in the Capture Current View wizard so that the defined code moves the model to the middle of the display, fills the display, but does not change the orientation of the model?

   a. **Save and Restore current view extents**

   b. **Don't change view size on restore**

   c. **Fit to model extents on restore**

7. Which of the following are valid measurement references when using the **Measure** functions in an iLogic rule? (Select all that apply.)

   a. Surface

   b. Edge

   c. Work Plane

   d. Sketch

8. Incorporating the following line of code into a rule prevents the model from updating until the rule has been executed

   iLogicVb.UpdateWhenDone = False

   a. True

   b. False

Command Summary

Button	Command	Location
	Add Rule	• **Ribbon:** *Manage* tab>iLogic panel • (*right-click in the Rules tab in the iLogic browser*)

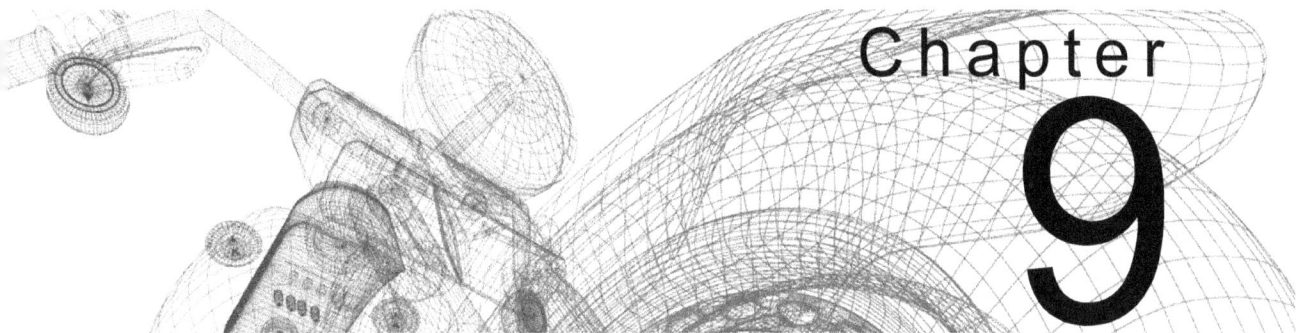

Project Practices

This chapter provides you with additional practice using iLogic functionality in part, assembly, and drawing documents. You are provided models to work with as well as a detailed list of design intent that is to be incorporated into the models. Detailed steps have not been provided for these projects. You are required to use your previous knowledge or return to and review the previous chapters, if required.

9.1 Introduction to Projects

Company ABC has seen a recent spike in custom product requests for the Mechanical Pencil design, shown in Figure 9–1. This product has become popular among potential clients to customize with their company branding and colors.

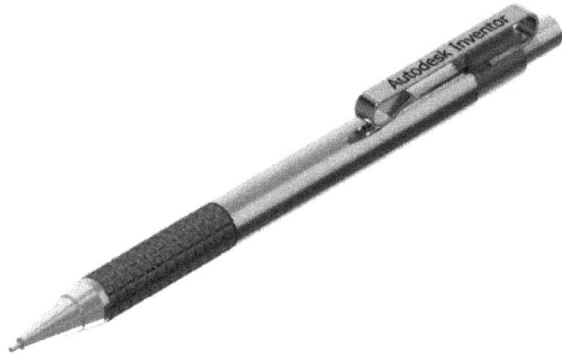

Figure 9–1

A need to reduce the engineering time required to model these requests and generate quotes has been identified. Typically, a quote acceptance is 30% and ABC company wants to increase this percentage. Having a stronger presentation of material and a shorter lead time to production helps with this goal. To this end, a need to create a configurator to better present the product to a prospective client has been identified. Once the quote is accepted, the files used in the sales configurator can easily be reused in engineering.

Practice 9a

Create a Configurable Model

In this project, you add iLogic rules to a Clip component that exists in a Mechanical Pencil assembly. The intent of the project is to incorporate a provided list of design criteria for the clip using iLogic rules. You will use the knowledge that you have gained throughout this student guide to add the required parameters, rules, and triggers to the model. One variation of the Clip component is shown in Figure 9–2.

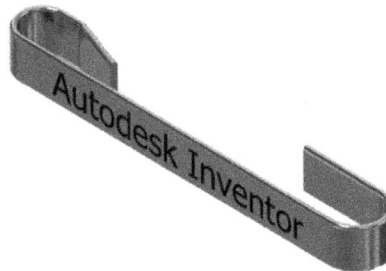

Figure 9–2

The Clip component in the model is not standard. This is the component to which the majority of configuration is done for the clients. This component is available in various widths, lengths, clip angles, colors, thicknesses, and can include custom engravings. When selecting an overall design approach, the use of iParts was ruled out due to the file system usage and the sheer number of variations that are required. Adding iLogic rules to the design is the approach to be used. The model will be saved as a template file and all the configurable parameters will be prompted for when the template is used to create a new model.

Sizing

The Clip has the following sizing criteria that must be met.

Model Parameter	Available Sizes and Notes
Width	• Standard sizes of 3.5, 4, 4.5, 5, 5.5 mm are available.
Clip_Angle	• Angular values between 7.5 and 20 deg can be defined.
	• Values outside of this range will not be accepted as they cannot be manufactured.
	• Set angular values over 20 to 20 deg and values under 7.5 to 7.5 deg.

Length	• Length values between 35 and 70mm can be defined.
	• For values outside of this range, provide a prompt that indicates the length is outside the accepted value.
	• Set values over 70 to 70 mm and values under 35 to 35 mm.
Thickness	• Standard sizes of .5, .6, .7, .85, 1 mm are available.
	• Alternate values are not available as these are the standard stock material thicknesses that are available.

Color

The Clip has the following color criteria that must be met. The material will stay set as Steel and only a color override is assigned. Create a user parameter called **Clip_Color** and set the options shown below as its possible values.

• The color of the clip can be assigned to one of the following.

Clip_Color Parameter Options	Color to be Assigned
Steel	Steel Blue
Gold	Gold - Metal
Silver	Silver
Gunmetal	Gunmetal - Polished
Titanium	Titanium - Polished

Engraved Text

The Clip allows for custom engraved text. The following design criteria must be met for the text.

• Engraved text can be included.

• User-defined entry for the engraved text is allowable, if required.

• The clip length determines the character limit on any custom engraved text. The character limits are 20 characters for 35 mm with an additional 2 characters available for every 5 mm of clip length. Notification must be presented for these limits when custom engraved text is being used. A notification warns of the limit but does not check the length of the engraved text.

Standard Clients

The following design criteria must be met to deal with quotes from our standard clients.

- Standard client names must be available from a selectable list (Paperfriend, Bik, Pentil, Co-Pilot, Sharpei).

- Selected client names must be exported to the model's iProperties. This will be used in the data management system for tracking and for the drawing borders.

Configurator

A local form is required to define all of the design criteria that has been discussed. Fields in the form will include selectable lists and user-entry fields for the Clip configuration. An image of the Clip model has been included in the practice files for use in the form, if required. The layout of the form is flexible but must allow for the required inputs. Setup the form to launch with an iTrigger or when the file is used as a template.

Open the model and design the configurator.

1. Open **clip.ipt** from the *Projects* folder. The model displays as shown in Figure 9–3.

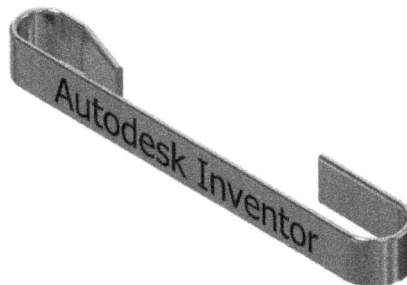

Figure 9–3

2. Review the model parameters in the Parameters dialog box. The key model parameters that are required in iLogic rules have been renamed for you.

3. Create user parameters that enable you to capture the design intent described above. The **Engraved_Text** user parameter has already been created and is currently being used in the Emboss feature.

   - Create a user parameter (**Client**) that enables you to select the client name from the defined list (**Paperfriend**, **Bik**, **Pentil**, **Co-Pilot**, and **Sharpei**).
   - Create a user parameter (**Clip_Color**) that enables you to select the clip color from the defined list (**Steel**, **Gold**, **Silver**, **Gunmetal**, and **Titanium**).

4. The **Client** user parameter must be converted to a custom iProperty using an iLogic rule. In the rule, assign the **Custom** snippet in the *iProperties* snippet category. Replace "PropertyName" with a new name for the custom iProperty (i.e., **Client**) and set the rule equal to the name of the **Client** user parameter.

5. Verify that the user parameter that assigns the client name is now listed in the Custom iProperties list for the model.

6. Define the required iLogic rules to capture the design criteria specified above.

7. Create a form that can be used to enter all the design criteria to allow for efficient configuration of custom designs.

*To save the drawing for use as a template, select **Save As>Save As Template** in the Application Menu.*

8. Set a trigger to execute the configured form when the model is used to create a new model. Additionally, set the form to trigger using an iTrigger so that it can be executed directly in the source model.

9. Save the source model. A completed model called **Configured_Clip.ipt** is provided in the *Projects* folder for you to review the completed iLogic code.

10. Test the configurator using both execution techniques.

Practice 9b | Create a Configurable Assembly

In this project, you add iLogic rules at the assembly level of the Mechanical Pencil. The intent of the project is to incorporate the provided design criteria in the assembly using iLogic rules. You will use the knowledge that you have gained throughout this student guide to add the required parameters, rules, and triggers to the model. The Mechanical Pencil assembly is shown in Figure 9–4.

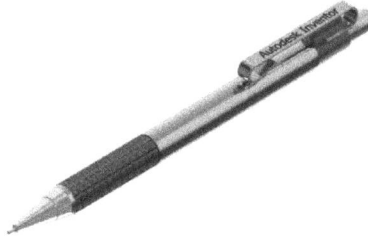

Figure 9–4

In the company's ongoing configuration project, you are expanding the use of iLogic rules to the Mechanical Pencil Assembly. The goal in doing this is to foster the interchange of components and selection of the Clip configuration at a higher level of the design. In doing this, all customization is accomplished from the assembly model.

The assembly must allow for part-level color change of specific components in the assembly and the interchangeability of the grip component with alternate design models. A form is required as an overall configurator tool. All of the Clip configuration options will be included in the assembly form and the assigned values must be pushed down to the model to update the geometry. The specifics of the design intent are described below.

Color

In conjunction with the color variations already driven in the Clip component, the Ring and Sleeve components must have the same color options as the Clip.Component Variations

Model	Colors to be Assigned
Ring	• Steel Blue, Gold - Metal, Silver, Gunmetal - Polished, Titanium - Polished
Sleeve	• Steel Blue, Gold - Metal, Silver, Gunmetal - Polished, Titanium - Polished

The Clip is configured at the part level, but the same component is always used during the quotation stage. Once a proposal is accepted, the redesigned clip will be saved out as its own unique component for use in a customer-specific assembly.

In the overall configuration of the assembly, the Grip component is varied using a different method. The Grip has two possible models variations. They are described as follows:

Component Name	Image
Configured_Grip_Style1.ipt (Note: Code to control the color change in this model has been created for you.)	
Configured_Grip_Style2.ipt (Note: Code to control the color change in this model has been created for you.)	

The two components described above are provided in the *Projects* folder. The current assembly uses a standard **Grip.ipt** component. In order to use iLogic to switch between the **Configured_Grip_Style1.ipt** and **Configured_Grip_Style2.ipt** components, you must replace the Grip component with both of these models. Both models are constrained in the assembly and the suppression of one or the other is controlled by the value of a user-defined parameter that defines the Grip Style type. In order to control the suppression with an iLogic rule, a Level of Detail representation must exist and be active.

The Grip must be of the same color as that specified for the Clip, Sleeve, and Ring. The rubber portion of the geometry has been assigned as black by overwriting the model color with a feature color. Edit the assembly rule that was created for the color of the Sleeve and Ring to ensure that the color of the two new grip components also reflect the selected color.

Configurator

A local form is required to define all of the design criteria that has been discussed. Fields in the form will include selectable lists and user-entry fields for the Clip configuration, the color assignment, and the Grip style selection. An image of the model has been included in the practice files for use in the form, if required. The layout of the form is flexible but must allow for the required inputs. Setup the form to launch with an iTrigger.

Open the model and design the configurator.

1. Open **Mechanical Pencil.iam** from the *Projects* folder. The model displays as shown in Figure 9–5. If you did not complete the Clip configuration in Project 1, open **Mechanical Pencil_Project1_Complete.iam**. This assembly references the **Configured_Clip.ipt** component with all the design intent programmed into the required iLogic rules.

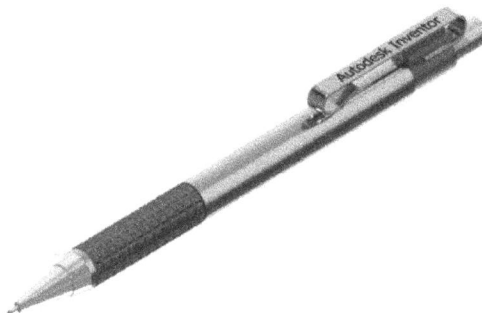

Figure 9–5

2. Create user parameters and define the required iLogic rules to capture the design criteria for Color and Grip style. The assembly must also accept entry of all the configurable Clip parameters (6); therefore, these user parameters must also exist at the assembly level. Once this is done, a rule must be established that pushes the parameter values from the assembly into the Clip component to initiate the change of geometry.

3. Create a user parameter that enables you to select the client name from a defined list (**Paperfriend**, **Bik**, **Pentil**, **Co-Pilot**, **Sharpei**).

4. The **Client** user parameter must be converted to a custom iProperty using an iLogic rule. In the rule, assign the **Custom** snippet in the *iProperties* snippet category. Replace "PropertyName" with a new name for the custom iProperty (i.e., **Client**) and set the rule equal to the name of the **Client** user parameter.

5. Verify that the user parameter is now listed in the Custom iProperties list for the assembly.

6. The client name parameter must also push the parameter value from the assembly into the Clip component.

7. Replace the Grip component with both **Configured_Grip_ Style1.ipt** and **Configured_Grip_Style2.ipt** components.

In order to suppress components in an assembly using an iLogic rule, LOD must be active.

8. Create a Level of Detail of Representation (i,e., iLogic) and activate it.

9. Create a rule to control the suppression of one or the other of the replaced grip styles based on the selection of a specific user parameter value.

10. Create a form that can be used to enter all the design criteria to allow for efficient configuration of custom assembly designs.

11. Set an iTrigger to execute the configured form.

12. Save the assembly. A completed model called **Mechanical Pencil_Project2_Complete.iam** is provided in the *Projects* folder for you to review the completed iLogic code.

13. Test the configurator using the **iTrigger** command.

Practice 9c	# Create a Configurable Drawing

In this project, you add iLogic rules to a drawing. The intent of the project is to provide with a design criteria for the drawing so that it can be captured using iLogic rules. You will use the knowledge that you have gained throughout this student guide to add the required parameters, rules, and triggers to the drawing.

One final configurator is required to document the assembly and clip design. A drawing template is required that enables you to start a new drawing that already has model views on it, and then be prompted to select the sheet size and drawing type. It must also correctly reference the drawing type parameter for use in the title block. A form is required to aid in the process of filling out drawing specific data.

Sheet Size

The custom drawing template must meet the following design criteria for its size.

- The Sheet Size selection must be based on the user input. The available sizes are (A, B, C, D).

- Ensure that tables or parts lists located on the top or right edges of the sheet are moved along with the change in sheet size.

Drawing Type

The custom drawing template must provide for user-selection of whether the drawing is to be used for Quoting purposes or for Manufacturing. The selected option must display in the drawing title block.

View Layout

The drawing must include an Isometric view of the assembly and views of the clip to document the client's configuration. Use iLogic code to scale and reposition the views for the various sheet sizes.

Configurator

A local form is required to define all of the design criteria that has been discussed for the drawing. Fields in the form will include selectable lists for the drawing size and type. The layout of the form is flexible but must allow for the required inputs. Setup the form to launch with an iTrigger or when the file is used as a template.

Open the model and design the configurator.

1. Open **Mechanical Pencil_Drawing.dwg** from the *Projects* folder. The drawing displays as shown in Figure 9–6.

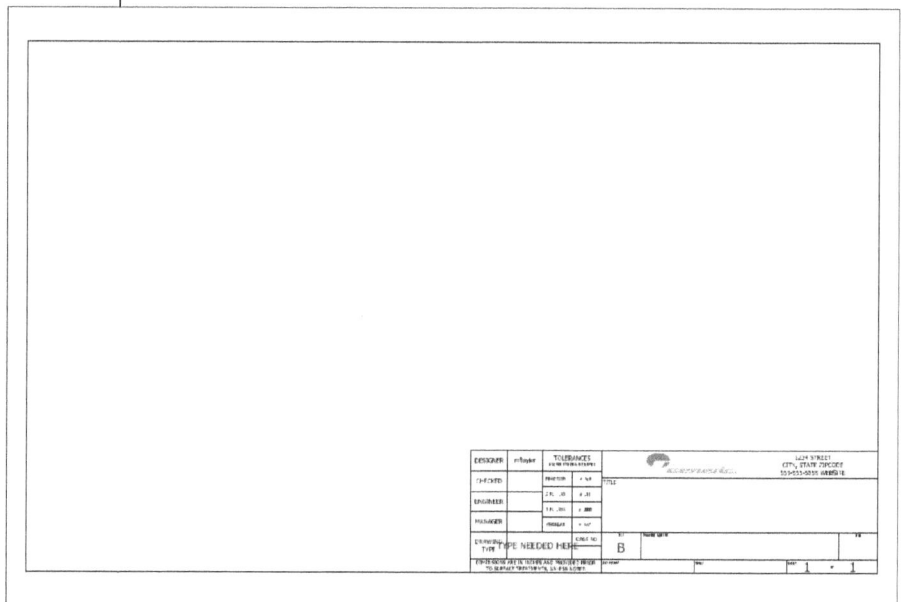

Figure 9–6

2. Create a user parameter to control the size of the drawing (A, B, C, or D) and write the drawing rule to control the assignments of borders to the drawing, based on the selection. Ensure the code controls the relocation of the title block with the change of size. The drawing should update to reflect the current sheet size value.

3. Create a user parameter that enables you to select the Drawing Type as either **Quote** or **Manufacturing**. This parameter (i.e., **DrawingType**) is required in the drawing's title block.

4. To incorporate a user parameter into the title block, you must convert the parameter (i.e., **DrawingType**) to an iProperty. This can be done using an iLogic rule. In a rule, assign the **Custom** snippet in the iProperties snippet category. Replace "PropertyName" with a new name for the custom iProperty (i.e., **DrawingType**) and set the rule equal to the name of the **DrawingType** user parameter.

5. Verify that the user parameter is now listed in the Custom iProperties list for the drawing.

6. Once you have created and converted the user parameter to an iProperty, edit the title block and replace the *TYPE NEEDED HERE* placeholder with the new custom iProperty.

7. Change the size and drawing type parameter values and ensure that the sheet size and title block update correctly. Set the size to **A** before continuing.

8. Add an Isometric view of the **Mechanical Pencil_Project2 _Complete.iam** (iLogic Level of Detail representation) to the top right corner of the drawing, similar to that shown in Figure 9–7. Create a custom orientation that uses the default Iso view in the model. Select a scale value that is suitable for the sheet size

9. Add a base and two projected views of the **Configured_Clip.ipt** component, similar to that shown in Figure 9–7.

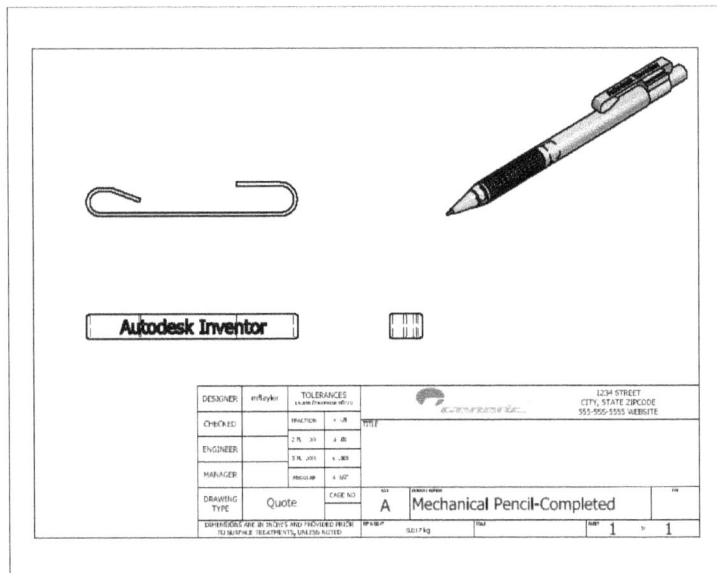

Figure 9–7

10. Save the drawing.

11. Edit the rule that was created to define the sheet sizes based on the user-parameter value. Add code to the rule that resizes and repositions all the views for each of the sheet sizes so that the layout is satisfactory.

12. Create a form in the drawing to prompt for the sheet size and the drawing type. Add the **OK**, **Cancel**, and **Apply** buttons to the form so that the drawing does not update after each parameter change. The update does not occur until after the form is closed using **OK**.

13. Set an iTrigger to execute the configured form. Additionally, set an event trigger so that if the drawing is used as a template, the form runs immediately when a new drawing is created using the template.

To save the drawing for use as a template, select Save As>Save As Template in the Application Menu.

14. Save the drawing to current working folder and as a drawing template. A completed drawing called **Mechanical Pencil_Drawing _Complete.dwg** is provided in the *Projects* folder for you to review the completed iLogic code.

15. Test the drawing configurator using the **iTrigger** command.

16. Test the part or assembly configurators to ensure that the drawing updates as expected.

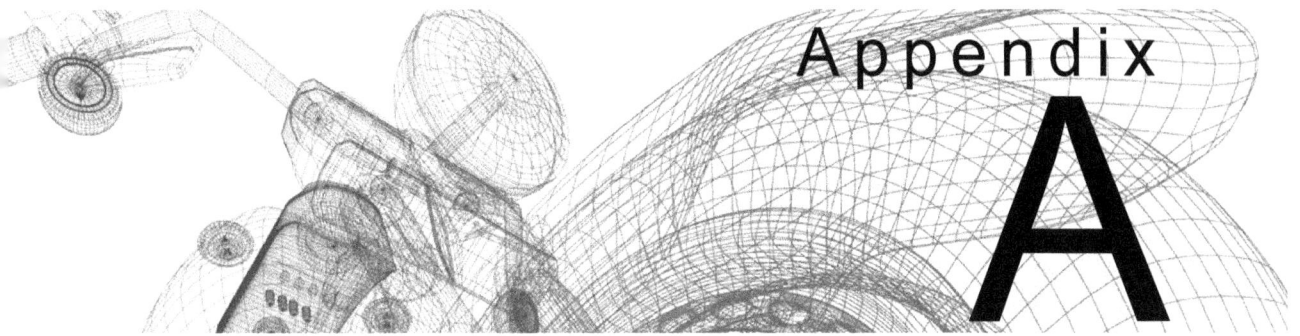

iLogic Design Copy

The iLogic Design Copy option provides a tool to create copies of a design. It can be used to strip a model of all its rules to provide a fully configured model or it can be used to create a newly named duplicate of the model with all rules maintained in the design.

Learning Objectives in this Appendix.

- Copy selected files so that iLogic rules are maintained in the copied versions.
- Copy a fully configured version of the selected files so that iLogic rules are removed in the copied versions.

A.1 iLogic Design Copy

The functionality to copy a design is also available in the Autodesk® Vault software; however, if this software is not installed, you can use the iLogic Design Copy functionality in your standard Autodesk® Inventor® software.

The **iLogic Design Copy** command, as its name implies, is used to copy an existing design to a new design. Although the command name includes the word *iLogic*, it does not mean that iLogic rules must exist in a model in order to use this functionality.

A benefit of the iLogic Design Copy functionality over the Copy Design functionality available in Autodesk Vault is that it takes into account rules that exist in the models being copied and enables you to determine whether those rules are copied or not.

The following describes the scenarios in which **iLogic Design Copy** is used in models that contain iLogic rules.

- The copied version of the model represents a fully configured design and does not need to retain the iLogic rules.

- The copied version of the model represents an exact duplicate of the original model that retains the iLogic rules for future configuration.

In either of these situations, the same iLogic Design Copy dialog box is used and different menu options are selected.

How To: Copy a Design Using the iLogic Design Copy command

1. Prepare the model by ensuring all rules exist and if copying a fully configured design, configure the model before continuing.
2. Save and close the model.
   - In order to execute the **iLogic Design Copy** command, all files in the Autodesk Inventor software must be closed.

3. In the *Tools* tab>iLogic panel, click ⬚ (iLogic Design Copy), as shown in Figure A–1.

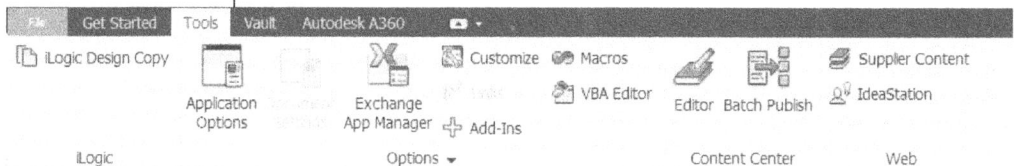

Figure A–1

Projects that are selected for copying must have a workspace definition. Documents outside the workspace are reused in the copied version, not copied.

4. In the iLogic Design Copy dialog box (shown in Figure A–2), define the Project in which the model resides.

   - The list of projects in the Source Project drop-down list is generated from the project files that are located in the *..\Users\username\Documents\Inventor* directory. To add a new project, if required, click **Projects** and use the Projects dialog box to add a new project.

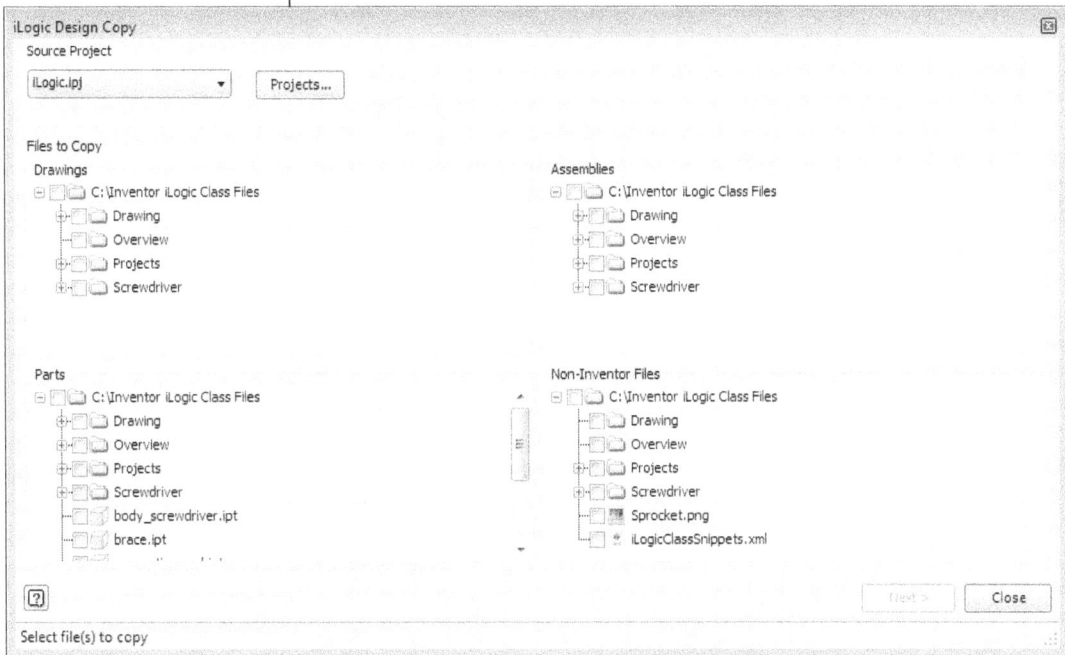

Figure A–2

5. In the *Files to Copy* area of the dialog box, select the files to be copied. To select, select the empty box adjacent to the filename to add a checkmark (☑).

   - Select the top-level project folder in any of the areas to select all files in the project.

   - In the *Drawings* area, you can select a drawing file to select it and any drawing models that are referenced in the drawing.

   - In the *Assemblies* area, you can select an assembly file to select it and any component models that are referenced in the assembly.

   - In the *Parts* area, you can select individual part models.

   - In the *Non-Inventor Files* area, you can select files that are associated with the project, but are not explicitly referenced (e.g., image files, spreadsheets, text documents, documents referenced in iLogic rules).

When copying assembly files, verify that the component names in the assembly are stabilized in the Model browser. This ensures that the rules referencing the copied components are to run correctly after suffix and prefix characters are added during the copy operation.

In the example shown in Figure A–3, the **Vise.idw** drawing is selected in the *Drawings* area. The referenced drawing models (parts and assemblies) are automatically selected. No additional *Non-Inventor Files* have been selected for copying.

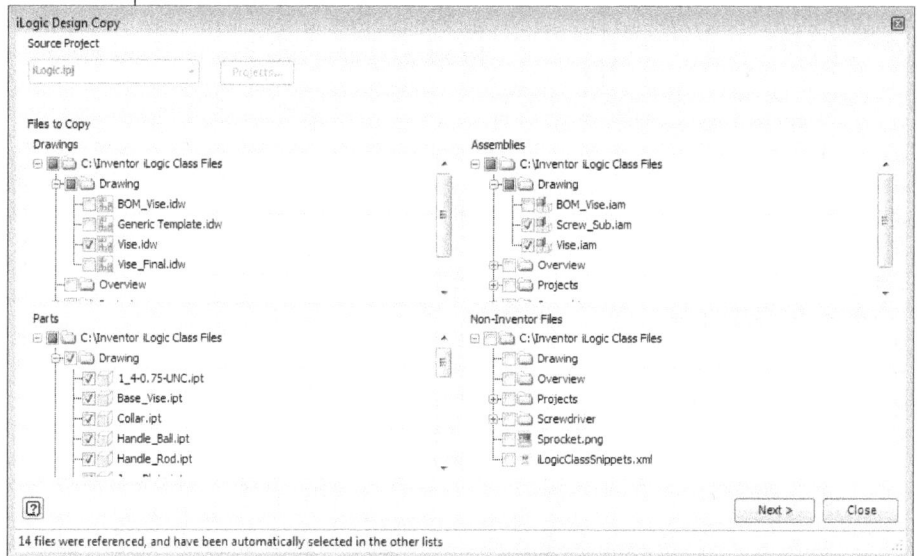

Figure A–3

6. Click **Next**. The iLogic Design Copy Settings dialog box opens, as shown in Figure A–4.

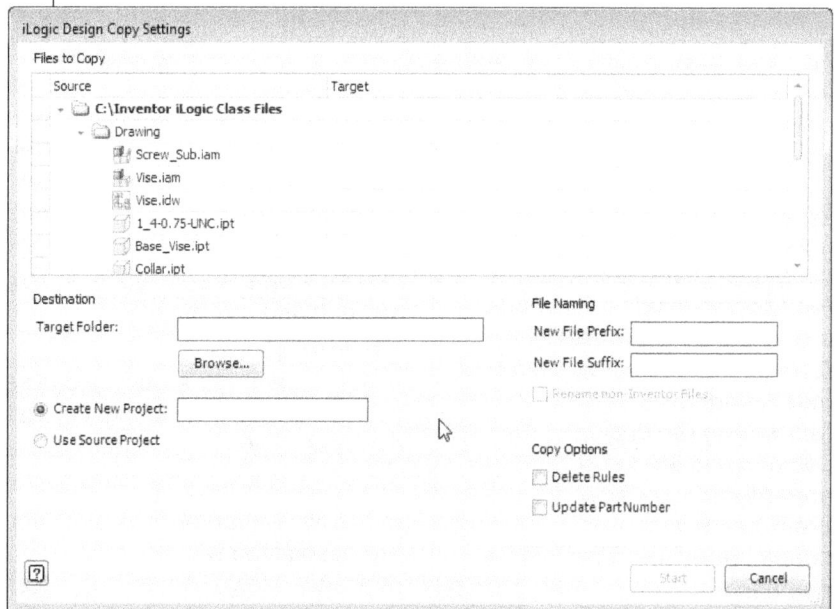

Figure A–4

The *Files to Copy* area lists all the files that are being copied. The lower portion of the dialog box provides the copy options. The options are described as follows:

Option	Description
Target Folder	Defines the location in which the selected files are to be copied. Click **Browse** to navigate and select a folder.
Create New Project	Specifies that a new project file is to be created and defines its name. All attributes from the original project file are maintained in the copied version.
Use Source Project	Enables you to use the source project file as the project for the newly created files.
New File Prefix/Suffix	Enables you to define prefix and suffix characters to be added to the names of the newly created files.
Rename non-Inventor Files	Enables you to define whether the files selected in the *Non-Inventor Files* area are renamed during the copy operation. It is recommended not to rename files that are referenced in iLogic rules; otherwise, rules need to be edited to run as expected.
Delete Rules	Enables you to determine whether the rules in the files selected for copy are maintained or deleted in the newly created files. For example, if a fully configured model is being created, the rules are not required and this option can be enabled.
Update Part Number	Enables you to specify whether the Part Number iProperty is to be updated in the newly created files.

7. Click **Start**. Once the copy process is complete, the Design Copy Progress dialog box opens providing details on the operation.
8. Click **Close** to close the Design Copy Progress dialog box.
9. Navigate to the new folder and review the newly created files.

Chapter Review Questions

1. The iLogic **Design Copy** option can only be used when there are rules in the model. If there are no rules you can use the copy techniques in the Autodesk Vault software.

   a. True

   b. False

2. Which of the following are true statements regarding the use of the iLogic **Design Copy** option? (Select all that apply.)

   a. New prefixes and suffixes can be added to the copied versions name.

   b. iLogic rules can be removed from the model.

   c. The iLogic **Design Copy** option must be executed with the file that is being copied open.

   d. Files that are children of the file being copied must also be individually selected.

3. The copied file must reside in the same folder as the target model?

   a. True

   b. False

Command Summary

Button	Command	Location
⬚	**iLogic Design Copy**	• **Ribbon:** *Tools* tab>iLogic panel

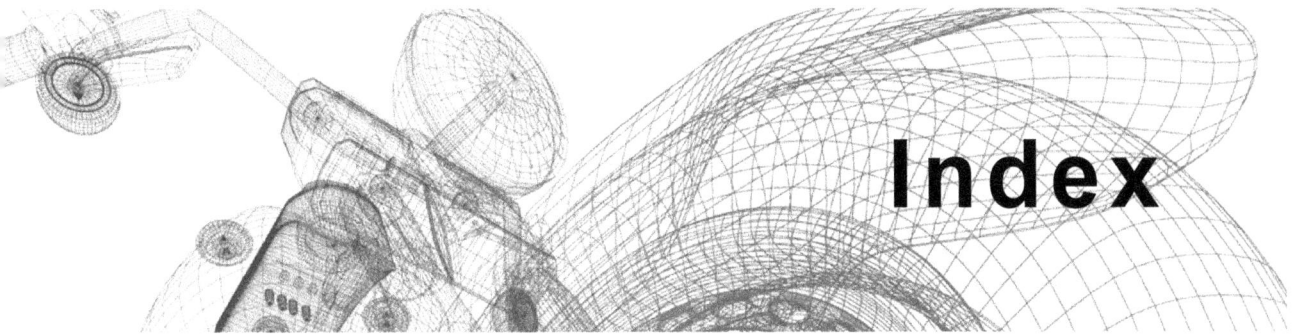

Index

www.ingramcontent.com/pod-product-compliance
Lightning Source LLC
Chambersburg PA
CBHW080934220326
41598CB00034B/5780